THE OMNI ADVANTAGE

Accelerating the Behavioural Change with Omnichannel in Pharma Sales Engagement

MEHRNAZ CAMPBELL

ONYX PUBLISHING

First published in 2024 by Onyx Publishing, an imprint of Notebook Group Limited, Arden House, Deepdale Business Park, Bakewell, Derbyshire, DE45 1GT.

www.onyxpublishing.com
ISBN: 9781913206536

Copyright © Mehrnaz Campbell 2024

The moral right of Mehrnaz Campbell to be identified as the author of this work has been asserted by her in accordance with the Copyright, Design and Patents Act, 1998.

All rights reserved. No part of this publication may be reproduced, stored in or introduced into a retrieval system, or transmitted, in any form or by any means (electronic, mechanical, photocopying, recording or otherwise), without the prior written permission of the publisher, aforementioned. Any person who is believed to have carried out any unauthorised act in relation to this publication may consequently be liable to criminal prosecution and civil claims for damages.

A CIP catalogue record for this book is available from the British Library.

Typeset by Onyx Publishing of Notebook Group Limited.

Praise For Mehrnaz Campbell and *The Omni Advantage*

In this book, Mehrnaz is perfectly placed to put herself in the shoes of a customer-facing field person and to understand the challenges they encounter when connecting remotely with customers. As a reader, you can 'feel' the opportunities and 'answer' any challenge you may encounter on your omnichannel journey. I was happy to be a 'sparring partner' for Mehrnaz, as she is one of a kind!

—KOEN JANSSENS
Global Associate Director, Field Force Effectiveness, Norgine

Distilling remarkable insights through her collaboration with industry leaders, Mehrnaz delivers here a brilliant 360 of the tangible actions we could and should take in order to improve the pharmaceutical industry's impact, for the benefit of HCPs and patients. A must-read for anyone who works in or with the industry and wants to contribute to a better, sustainable, and human healthcare.

—FLORENT EDOUARD
SVP, Global Head of Commercial Excellence, the Grünenthal Group

The Omni Advantage cuts to the core of why, despite all the recent technological advances and digitalisation driven by the pandemic, pharma has not yet reached omnichannel nirvana—and what the industry can do to get there.

—PAUL TUNNAH
Founder, pharmaphorum & Chief Content Officer and MD, Healthware Group

Mehrnaz's *Omni Advantage* is authentic, riveting, and straight from the heart. Mehrnaz's brutal honesty in the narrative, with a personal touch and practical actions, will make you go from, 'Why omnichannel?' to, 'Why not omnichannel?' as an empowered marketeer, sales rep, commercial leader, or someone driving digital change in their organisation. It truly puts you at an advantage to read this book!

—VANITHA VENKATASUBRAMANIAM
Global Customer Engagement Strategy Lead, Novo Nordisk

You've spent a fortune, committed thousands of hours, and made outlandish promises. You press the big red button and... nothing. Tumbleweed. Crickets. Embarrassment! It sounds trite, but it happens every day. Our industry is famous for investing in the next shiny toy but failing to see results. Mehrnaz has seen it, too—but instead of endlessly tinkering with the technology, she comes back to the people, who we know will remain the engine of growth even in an era fuelled by AI. Read her book if you want to stay ahead!

—PAUL SIMMS
Chief Executive, Impatient Health

To date, all the books that we have read on this topic have been theory, and Mehrnaz brings the practicality of how you do this in pharma.

—GAURAV SANGANEE
MD and Founder, Closing Delta

Due to a shortage of staff and a backlog of patients post-COVID, many health services are operating under immense pressure, so it has never been more important to enable and empower our field teams to be an effective part of the omnichannel outreach and add true value at every HCP engagement. In *The Omni Advantage*, Mehrnaz has used her extensive knowledge and years of experience to create a thoughtful step by step guide that makes core reading for anyone wanting to achieve excellence in this space.

—JAMES HARPER
Founder and Managing Director, twentyeightb

Pharma still operates in silos today, and whilst we all agree that this needs to change, we are struggling to make it happen. This book goes deep into the intricate dynamics between sales, marketing, and medical, driving you to think about the human elements that come into play when trying to effect change. I thoroughly enjoyed reading this book and have taken away not only a fresh perspective on certain scenarios, but also some tangible steps that I can apply today.

—AMISH PATEL
Marketing Consultant, APC

I dedicate this book to my parents, who taught me resilience and encouraged me to use my head, above all else.

Contents

Foreword ... 11
Preface ... 15

Part I: How Did We Get Here?

Chapter 1: Oops! Wait A Minute, Have We Forgotten Something? ... 23
Chapter 2: The Squeezed Middle ... 34
Chapter 3: Why Isn't the Horse Drinking? ... 47
Chapter 4: HCPs Are Human, Too! ... 69
Chapter 5: Raising the Tide Lifts All Boats ... 81

Part II: How Do We Get Out of Here?

Chapter 6: Shifting Gears ... 95
Chapter 7: The Human Transformation of Digital Engagement: How Did We Do It? ... 106
Chapter 8: From Message Bearers to Trusted Advisors ... 124
Chapter 9: The Future Field Teams ... 136
Chapter 10: The 'Doing' Chapter ... 144

Postface ... 148
Acknowledgements ... 153
About the Author ... 155
Resources ... 156

Foreword

Paul Simms, Chief Executive at Impatient Health

Children starting school this year will be retiring in 2065. Nobody has a clue what the world will look like in five years' time, let alone in 2065. And yet, we're meant to be educating them for it.

Picasso said that all children are born artists. The problem is to remain an artist as we grow up. We don't grow into creativity, we grow out of it... or rather, we get educated out of it.

—Sir Ken Robinson, February 2006

In May 2023, the World Health Organization declared that the COVID-19 emergency had ended. *Phew*. Officially, we could now place all our attention on the future.

But hang on. Look out of the window. The frenzied race of AI chatbots, the inflationary economic surge spurred by the unending conflict in Ukraine, and the deepening political chasm between red and blue or east and west (take your pick!), collectively brew a potent cocktail of uncertainty for our foreseeable future. The one constant we can be sure of is change.

Although our digital world forces us to obsess over tools, channels, and processes, the truth is that it's not tools that will define us. Tools will come and go. The only way to futureproof ourselves is through a wholesale change of attitude; a desire to be flexible, or even experimental, in our approach. Only then can we have the confidence to say that no matter what surprises the world tries to throw at us (and we can be sure that there are plenty on the way), we will prevail.

What does that mean for us executives in the life sciences industry? Very simply, it means infusing more science into non-R&D roles. I don't mean more boffins and brainiacs; I mean nurturing curiosity, objectivity, and lack of assumption—but without dimming the flames of passion.

This invaluable book provides a plethora of practical tips you can apply to your work come Monday morning, but the sum total of all the nuggets spread across its chapters is an upgrade in your attitude, confidence, and willingness to take on the world in its myriad forms. Instead of shying away from the scientific method, you'll embrace it.

The turmoil of the last few years has left a daunting backlog on many of us. Most countries are grappling with huge hospital waiting lists, with medical appointments lagging months (or even years) behind schedule. Strikes and protests compound these delays, and this means that millions of diseased patients languish undiagnosed. Many will discover their fate with too few minutes left to reverse it.

Against this backdrop, how can we justify that the HCP spends time with us? How can we rationalise more meetings with the commercial interests of the pharmaceutical industry?

The answer to this question is simpler than it seems. Again, we have to look within ourselves. We need to adjust our own mindset away from one that says, 'Sell whatever you can today,' to one where we add value with every interaction; where we ensure every interaction empowers physicians to excel in their practice; where we cultivate greater trust through impact.

As this book attests, our industry has predicted the demise of the pharma field professional (or the MSL) many times before, but the truth is that even in the aftermath of the pandemic, field teams are bouncing back—but in a different mould from before.

Our duty now is to ensure that the field professional assumes the right role—that is, a partnership role that is integral, beneficial, and value driven. Mehrnaz and her contributors aptly describe this as the 'Omni Advantage'. I implore you to use this book as a trusty companion as you shape your customer interactions going forward.

As the late Sir Ken Robinson wisely reminds us, planning for a future we can't comprehend is an exercise in futility. All we can do is adapt ourselves, be as prepared as possible, and embrace (and even enjoy) the unpredictable landscape ahead.

Preface

As I flew over the handlebars of my bicycle and the asphalt came to meet my shoulder with a thud, I was determined to make sure the last two hundred and ninety miles of pain, sweat, and tears had not been endured for nothing. I was going to cross the finish line, come hell or high water.

It was a sunny day in May 2016, just outside Paris. I was on a charity bike ride, cycling from London to Paris with my colleagues to raise awareness for myeloma, a malignant tumour that grows in the white blood cells in bone marrow.

I was ten miles away from Paris when I fell off my bike and broke my shoulder and upper arm in several places. I was more annoyed that I couldn't finish the ride than I was about my injuries.

The team loaded my bike on the back of the support van, and we drove to Paris behind the pack. My arm was throbbing, but I was blocking the pain out as I tried to formulate a solution. I was determined to cross the finish line on my bike, not in a van.

When we got to Paris, I insisted they stop the van to allow me to ride my bike. I was in no state to lift my right arm, let alone apply the brakes, so I used my left arm to lift my right arm on the handlebars and, flanked by two ride captains (who now had the job of giving my cycling shirt a tug to slow me down when necessary), I pedalled across the finish line. I joined my colleagues for a group photo in front of the Eiffel Tower and celebrated with a beer before heading to the hospital.

Upon my return to Edinburgh, I was signed off work for two months. At the time, I was working as a Regional Account Director (RAD) and needed to get to customer-facing meetings—but with my injuries, I was not insured to drive a car.

After a week of going stir crazy, I called my manager and persuaded him to let me return to work. We were negotiating an important commercial agreement with National Procurement Scotland, and I suggested that they allow me to work from home to start with. I argued that even though my arm was broken, I could use my brain and reach customers via phone and email.

They agreed and, during those coming weeks, I discovered something unexpected: because I wasn't driving, I had more time during the workday to dig deeper, read documents, listen to HCP feedback, and analyse the information. In the process, I picked up on a unique piece of information that could give us a competitive advantage. I shared this information with my medical colleagues, who used it as supporting evidence that changed the market dynamics of the tender process in our favour. Subsequently, our brand was selected as the recommended first-line choice across all three cancer networks in Scotland and in fourteen local formularies. By utilising multichannel marketing initiatives, we accelerated growth so much so that our brand became the UK-wide market leader within eighteen months.

This was my introduction to remote working.

Six months after the bike accident, I was planning to move to the US, and approached my employers with a business proposition: would they consider using me as a contractor so that I could continue working with them remotely from the US? They did not hesitate in saying yes. They said if I could achieve what I had from my sickbed in Edinburgh, there was no doubt that I could influence the business remotely from anywhere.

This remote working journey took me in a new direction: I became an independent service provider for pharma. I established Cheemia (my company), which I have been managing remotely from the US since April 2017.

Cheemia was set up initially to provide remote sales and marketing strategy advice and implementation to UK pharma companies who wanted to gain market access and grow their brands in Scotland. However, because our clients liked our approach so much, the service rapidly expanded across the UK. Our approach was unique in the fact that we didn't help pharma companies sell to the NHS; we helped the NHS to buy from pharma companies. Consequently, we grew (and still grow) sales exponentially.

When the pandemic happened, everyone was trying to figure out this new way of working, but we at Cheemia had not only figured it out, but were exceptionally good at it. Specifically, everybody in pharma was struggling to engage with HCPs, while we were able to continue doubling the sales growth of a brand for our client. The brand in question was a respiratory medicine, which was even tougher to engage HCPs in during COVID, but we still made it work.

We felt (and still do feel) that, as a company that had by that point already operated remotely for some time, we'd had a head start. We'd started our remote journey way before the pandemic, and we already had a recipe that was resonating with healthcare professionals. So, we felt that we had a duty of care to share our recipe with the rest of the industry, and we did that through Cheemia ReSET and Cheemia ReINSPIRE, an online training platform to help fellow pharmaceutical field professionals who had seen remote selling become an essential part of creating better omnichannel relationships with healthcare professionals.

In summer 2020, I was introduced to Koen Janssens, Global Associate Director of Field Force Effectiveness at Norgine, by Kym, an ex-colleague at Takeda, who was leading the global digital division. Koen was actively looking for a solution for this now-ubiquitous problem, and Kym, knowing this, put us in touch with each other. I invited Koen to jump on our beta testing and experience the platform for himself.

Norgine initially piloted it in four countries, and then they rolled it out across their global organisation, across twelve countries. I'd like to take this moment to note some stats that they shared with us, which were extracted from Veeva CRM. These stats depict the interactions that took place in their sales team in Belgium. What we wanted to examine was what was normal in 2019 (before the pandemic), what happened during the pandemic, and what happened after the team's adoption of this new recipe (in Q4 of 2020).

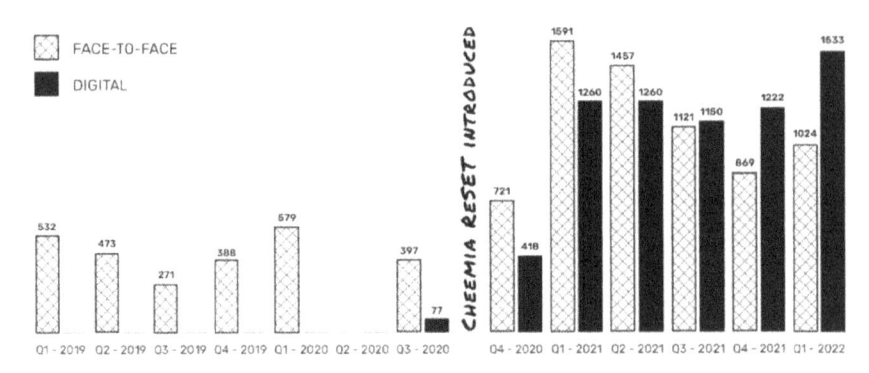

Example of a District Sales Team's Increased
Omnichannel Engagement After Cheemia ReSET

As you can see in the chart, before the pandemic, Norgine were engaging in about five hundred face-to-face interactions per quarter. Then, in 2020, with lockdowns in place across the globe, this number obviously fell to zero. But then, in Q3, face-to-face interactions started taking place again, and digital tools started to emerge. In Q4 (when the team started using our model), we saw their digital interactions start to increase. This we had expected. What we *hadn't* expected was for their face-to-face activity to increase dramatically as well—and that's exactly what happened. If you combine the team's digital and face-to-face interactions in Q1 of 2021, they're almost five times what they were pre-pandemic. Not only that, but their confidence in engaging with healthcare professionals had increased both in virtual interactions *and* in face-to-face meetings—a very unexpected byproduct of our service.

What we also find interesting is the fact that, if you compare the engagements twelve months after the team adopted our strategy (which brings us to Q1 of 2022) to the engagements they had *before* they adopted our strategy (in Q1 in 2019), they are still *five times* what they were before. This is pretty remarkable. Usually, when you do traditional sales training, you initially see a noticeable shift in behaviour, but people gradually revert to how they were behaving before. However, this did not happen in this example: with Cheemia ReSET, the shift in behaviour was sustained over time. The field teams quickly gained confidence in using the video channels and engaging with HCPs remotely across the whole global organisation. In parallel, as the field became more comfortable with use of digital tools and content, the demand for more digital content creation was initiated by the field force, not marketing.

In summary, 2021 was the best year Norgine had ever had, with the highest profit and turnover ever and all targets smashed. Even 2022 was a good year for growth, despite the economic crisis and changes with reimbursement.

What did we gain from these extraordinary results? Through our success with Norgine, they introduced us to Veeva Systems, who invited us to become an Emerging Veeva Services Alliances Partner. We also received a flurry of awards: Cheemia was named Top Remote Work Tech Solution Providers in 2021 by CIO Applications Europe, and in 2022, we were one of the CanDo Innovation Summit #SMEstage winners in the UK. We were also named a Scotland finalist in the Digital/e-Commerce Business of the Year Awards by the Federation of Small Businesses. I was also a finalist at the AccelerateHer Awards in the Disruptive Innovation category, which is supported by Barclays Eagle Labs.

'Disruptive innovation' was exactly how we changed field force behaviour in six weeks. Here, in this book, I want to share what we learned and what you need to do to achieve similar results in your field teams.

PART I

HOW DID WE GET HERE?

1
Oops! Wait A Minute, Have We Forgotten Something?

It is no secret that the pharmaceutical industry went through a massive change during and after COVID. Traditionally, the main channel for communication between pharma companies and healthcare professionals was face-to-face interactions, but COVID completely changed that, and, in turn, the whole HCP engagement model transformed overnight.

Many thought the change was temporary and everything would go back to normal at some point. Now, we know better. The industry has changed for good and will never return to how it was before, that's for sure.

Historically, the pharma industry has been slow in adopting technological tools compared to other industries. Before the pandemic, some pharma companies dabbled in e-detailing, email campaigns, and interactive websites, and most at least had digital content in PDF format. But as an industry, pharma remained very slow in digital transformation and omnichannel engagement. This prevented the industry from having a two-way omnichannel dialogue with HCPs. Some argue that the pharma industry's requirements pertaining to compliance, pharmacovigilance, and regulations were the main reasons for pharma being so 'behind the times' with digital technology. Regardless, pharma only truly embraced digital transformation once it had no choice but to do so; that's clear enough.

Basically, COVID was not just an accelerator; it was a long-overdue kick up the backside.

Accordingly, the change curve was fast and furious. Overnight, pharma had to adapt to remote working across all functions. This change affected some divisions and countries more than others. Some of the least affected were those who'd been working in the office, as those people continued to do the same work as before, albeit from home. Of course, working from home had its challenges, such as adjusting to co-sharing space with other family members and home-schooling, but this transition is incomparable to what those who were most affected—amongst them, field professionals—went through. For field professionals, the very nature of their role fundamentally changed overnight.

The change curve was steeper for field professionals because it was simply not business as they knew it. They weren't used to working in front of a computer all day, whether in an office or otherwise. On the contrary, they'd previously spent their working day travelling to various locations and talking to multiple HCPs in different roles and functions. They most certainly had not had a nine-to-five office-based job. In fact, they *hated* the idea of having a nine-to-five office-based job, so much so that they'd decided to pursue a career in this field specifically for that very reason: this field promised a dynamic, people-facing role.

In 2020, we at Cheemia did an in-depth survey with field professionals to test their preferences in this regard, and the participants expressed that, prior to the pandemic, they'd had variety, movement, and human interactions in their role; in contrast, after the pandemic, they had to stay at home, and this did not feel like going to work. They felt stuck.

During and after the pandemic, pharma companies rushed to undergo digital transformation (it was that kick up the backside I mentioned earlier). They invested in tech, upgraded their CRM systems to enable video calling, bought video platforms (such as Zoom and Microsoft Teams), and accelerated the development of digital content to (in theory) enable their field forces to engage with HCPs just like they had before. Most companies invested in training about Zoom, Microsoft Teams, remote working resilience, and managing an at-home working environment, but this was done as one-off group coaching. There was nothing disruptive or innovative about the style of the training. It was the same face-to-face training the field forces had always received; the only difference was that it had been delivered via Zoom or Teams instead.

The 'training' box had been ticked (or so they thought!).

Before long, senior leaders in pharma companies started questioning the field teams' output. The field professionals' access to HCPs was not at the expected level. This was, to the senior leaders, a baffling discovery. In their minds, they had invested in the needed technology and given their field professionals everything they needed for them to do their job just as effectively as before—so, if the field professionals were struggling to gain access to the HCPs, was the problem with the field force? Was the field force even needed anymore, if they were suddenly so ineffective? Likely, they concluded that the field team was not able to adopt the digital tools, and so they questioned whether their field forces were, in fact, as indispensable as they'd thought.

When I moved to the UK as a foreign student back in 1985, I was studying English at St Stephen's College in Kent. I thought it would be fun to do another subject—something creative that did not require me to speak English—and art seemed a natural choice. I was not the artist in the family: my sister was the talented painter. I always loved science and sports, and painting was not something I had ever tried before. Doing an Art O-Level whilst studying English just seemed a fun way to pass the time. My art teacher, however, didn't share this sentiment. She said, 'Don't bother doing O-Level Art. You won't pass.' She didn't give me any encouragement, and spent most of her time coaching others she deemed talented.

Undeterred, I turned up every week and practiced. Despite my teacher's negative expectations, I remained optimistic. She saw me as someone who was destined to fail, and I saw myself as someone who was determined to have a go. Passing the art exam was not going to change my life, but it gave me purpose and a target to aim for.

She was astonished when I got a B!

I didn't practice art again until I moved to the US a few decades later. We were visiting the Philadelphia Museum of Art when I saw an exhibition about watercolours. I watched a video demo by an artist, and suddenly felt a strong desire to have a go at painting with watercolours. This filled me with excitement.

The following week, I found a local art workshop and met a lovely teacher called Sue who specialised in watercolours. She was kind, patient, and positively optimistic about my ability to paint. I did not have many lessons, but was very surprised to find I could actually paint very well!

I guess the moral of the story here is that we all have natural abilities that may remain dormant if we don't have a go. Plus, having an environment and coaches that have positive expectations increases the likelihood of our success. So, if you are a pharmaceutical leader who is frustrated about your field force not embracing the digital tools and channels you have given them, you are not alone. Many leaders in pharma are feeling this way, and have actually started speaking about their frustrations openly. I have talked to many commercial leaders, field force effectiveness and customer excellence leads, and those involved in L&D, and they are all experiencing the same challenge: they invested in the required technology and digital content, and yet that technology and content is not being used—and they can't understand why. They're asking, *How do we get value from this tech investment? How do we engage the field professionals so they will adopt these remote channels and digital content? How do we get to the point where the field force can feel confident in engaging healthcare professionals using multiple channels? When will we see the impact of that improvement on engagement levels, the quality of our relationships with HCPs, and our sales figures?*

The root cause is simple: in the rush to invest in the technology, they forgot about the people, and the fact that people, as a fundamental part of their nature, resist change!

Digital transformation affects different people in different ways. Some people embrace new things and are quite excited by them, while some resist change, especially if they feel threatened by it. Perhaps because the commercial and marketing leaders in the office had a shared excitement about these new tools, they were expecting the field force to react in the same way as them: with curiosity and an eagerness to absorb it all.

If you are looking to get the maximum return on the investment you have made in digital tools, CRM systems, AI, and the high-quality digital content your marketing teams have developed, and you want to make sure your field professionals get their heads around it all, this is the book for you.

This book is designed to take you along this journey, showing you the key areas you need to address and introducing you to the practical tools and examples that will allow you to support your field force's mindset-based transformation. This will facilitate them confidently adopting the digital tools and content you have provided and, ultimately, creating better relationships with HCPs.

I'm not only going to give you the understanding and knowledge to do this, but I am also going to encourage you to roll up your sleeves, take massive imperfect action, and do something about the challenges you're facing.

Checkpoint 1A: Action Steps

- Ensure you have the appropriate technological tools in your company and available for your field teams' use.
- Ensure you are creating the right content in your company and that it is available for your field teams' use.
- Understand that your job here is not done: you need to get your people on board.
- Understand that whether you think your field team can or they cannot, either way, you are right. So, be patient, empathetic, and positively optimistic about your field teams' ability to embrace digital content and tools.

The Omni Advantage

*

To summarise the problem with which we are currently dealing: pharma got distracted by the shiny things. Specifically, pharma's leaders got distracted by the possibilities that came with digital channels and technology, and in turn, they forgot to focus on the human element of digital transformation. Specifically, they forgot to focus on the field force itself.

I have spoken to various commercial and marketing leaders in pharma who have shared that their first reaction to 'digital transformation' (i.e., the pandemic) involved buying the relevant tech and platforms. Buy the platform, job done. The next step (as they saw it) was to create digital content (modular content), and this was all very exciting because, suddenly, they were faced with multiple channel options for communication. In the past, the field force had been the main channel of interaction with HCPs, and so the adoption of technology, the improvement in tools, and the need to reach HCPs remotely led to the exploration of alternative channels, including those enabled by AI.

All these shiny new things were, and still are, exciting, and with various digital and creative departments and agencies constantly presenting new ideas to encourage marketing teams to pilot and invest in them, it was all very stimulating and had us pondering the possibilities: How do we incorporate AI into communication with healthcare professionals? How do we use AI to track the quality of these interactions? How do we create digital sales funnels to automate content delivery and HCP engagement? How do we use social media channels for listening and shaping behaviours? How do we utilise ChatGPT in our communications?

It's no wonder the shiny stuff distracted everybody! Yet the adoption of tech solutions requires a significant time investment. It requires the creation of a new workflow and training on how to deal with the large amount of data that it creates.

This led to commercial leaders and marketers using their time to focus on learning how to use and process this information, review leadership dashboards to monitor the digital data and insights, and check the click and open rates. These things essentially became new shiny metrics for leaders and marketers. This led to an array of new buzzwords entering the pharma vocabulary, including 'digital transformation', 'Net Promoter Score' (NPS), 'multichannel', and 'omnichannel', to name a few.

Interestingly, everyone uses these words daily, but I'm unsure whether 'omnichannel' means the same thing to everyone, or whether the difference between 'multichannel' and 'omnichannel' is clearly understood. During our Digital Transformation Series of interviews in 2022 and 2023, I asked European and North American pharma thought leaders to share their understanding of these terms through social media, and I was intrigued to hear their take on omnichannel, digital transformation, and the human element of digital transformation. The views and definitions varied widely.

In my company, we refer to 'omnichannel' as a strategic concept that seeks to provide a seamless, integrated experience across all channels of interaction with customers, whilst being an integrated way for cross-functional teams across the business to plan, execute, and respond. The illustration below clearly shows the difference between multichannel and omnichannel HCP engagement. In multichannel, all the channels are available to the HCP, but they aren't integrated. In omnichannel, on the other hand, all the channels that are available to the customer are

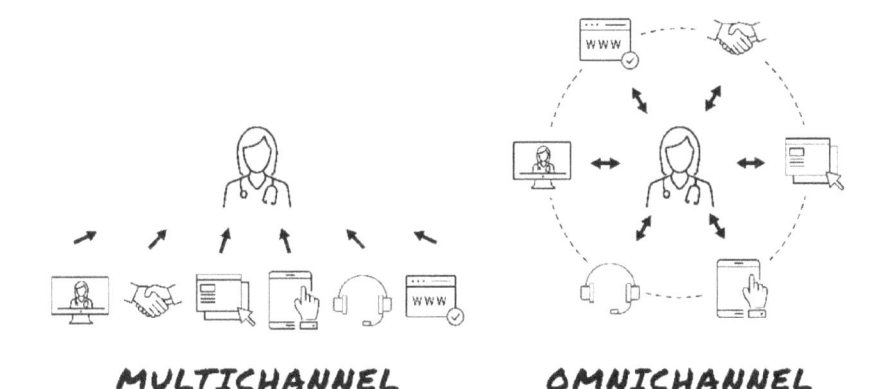

MULTICHANNEL   OMNICHANNEL

integrated. This means that although pharma talks about omnichannel communication, without the integration of channels, it is, in reality, multichannel communication.

So much time and focus is given to the debate around omnichannel and multichannel, but hardly anyone is asking why we are focusing so much on the channel itself. I was moderating a session at the NEXT Pharma Summit in Dubrovnik in 2023 and, during the panel discussion, I invited the audience of two hundred pharmaceutical leaders to stand up. I asked them to evaluate their company's content in terms of 'relevance', and to remain standing if they honestly believed their content was relevant. Half the room sat down.

Then, I asked those who remained standing to reflect on the timing of their communications, and to remain standing if they honestly believed their relevant content was being delivered to HCPs precisely when the HCPs needed it most. To my astonishment, they all sat down.

I'd been planning to ask them a question about their channel choice, but that now seemed redundant. If the content shared is not relevant and is not communicated at the time that HCPs need it most, then channel choice is (at least for now) completely immaterial.

Furthermore, with the radical increase in technology and remote working that arose with COVID, there came a huge increase in the volume of online communication. Inboxes were rapidly filled with emails, and the expectation to visit various platforms and be up to date with the ever-changing information affected everyone, whether you were in sales, marketing, medical, pharmacovigilance, or leadership. No matter the department you worked in (and no matter how easily a given employee adapted to this digital transformation), everybody became overloaded with digital communications and ended up working longer hours.

I can personally relate to this as, back in 2020, a new digital lead joined Cheemia to support our digital transformation and help us to improve and automate our processes. Prior to him joining us, we'd been using email as our main channel of communication and SharePoint for document sharing and storage. Then, he introduced Slack for internal communications and

Asana for project management. In time, I realised I was spending a big chunk of my day checking Slack, Asana, my emails, and my social media platforms. I stopped and asked myself, *Am I spending my time doing stuff that adds value?* The reality was that I was spending more time checking platforms than focusing on actually adding value.

I know I was not alone in doing this, especially during COVID. To some extent, we all ended up working much longer hours and being distracted by tech and the processes related to tech.

In other words: technology and ease of communication has, counterintuitively, increased our workload, even though it is designed to *simplify* our lives.

This reminds me of the M25, a circular motorway around London, which was designed to reduce traffic congestion. Before the M25 was built, there were traffic jams in London, and after the M25 was built… well, there were still traffic jams. It was just that now, there were *extra* traffic jams—on the M25. The facts are, even if you add another lane to the M25, you are still going to have traffic. When you create a channel, you create the traffic jam to go with it.

I'm not saying tech is bad; far from it. I have learned to use technology to simplify my processes and to help me connect with people. I'm embracing it, because it is necessary in today's world. At the end of the day, it saves me time and money and allows me to engage with HCPs in the UK whilst living in the US. Yet being distracted by the shiny things put emphasis on the technology itself rather than the main goal behind us *using* this technology: engagement. Pharma has retained old (outdated) systems and traditional ways of thinking, but with the tech bolted onto them. The new technological tools are being used to do more of the same of what the industry was doing before.

We need to shift our mindset from one where we push outbound communication about brands, to one where we use technology to create two-way communication with HCPs to help us to better understand their pain points and deliver value through their preferred channels. Specifically, we need to adopt the mindset that digital tools, platforms, and tech are *enablers* for HCP engagement, not the end goal.

Thus (and this is where we get to the crux of the issue), as an industry, we need to figure out how to engage and motivate the whole organisation to understand the value of the digital tools and why we need them: because they will benefit the future of our industry and make us more customer centric.

So, it is not just about *having* the platform, channels, and content. These are just the tools. It is about thinking about how to best *use* these tools to engage HCPs and to improve the field team's experience with using these tools. We can do this by looking at the tools from the field professionals' perspectives, in consideration of their personas and preferences, and taking them along this journey with us.

Some may argue we need tech savvy field professionals if we, as an industry, are going to use this digital content successfully. But the question is, where are you going to find these tech savvy field professionals? Even if they are out there, there is going to be huge competition between pharmaceutical companies to attract them—and if you do successfully bring them in, they are going to have to start from scratch building relationships with healthcare professionals.

We need to turn our attention to the 'people'-people we already have within our ranks who already have relationships with HCPs. We need to focus not only on getting them to be confident with using technology, but also on engaging their hearts and minds; on showing them that technology can improve the relationships they already have. As Quentin Descat, the global cardiovascular commercial launch brands lead at Bayer, said on a LinkedIn post, 'As long as doctors are human, field customer representation will be human.'

Checkpoint 1B: Action Steps

- Understand that being distracted by the shiny things puts emphasis on the technology itself, rather than the main goal behind us *using*

this technology: engagement.
- Embrace the mindset that digital tools, platforms, and tech are *enablers* for HCP engagement, not the end goal.
- If HCPs and field teams are not engaging with your content, check to see if your content is actually 'relevant'. Relevant content is that which directly addresses a pain point that HCPs are facing. Think 'relevance' first, and then once this is right, focus on your timing and your channel.
- Focus on people!
- Improve the field team's experience by looking at the tools from the field professionals' perspectives, in consideration of their personas and preferences, and taking them along this journey with you.

2

The Squeezed Middle

As our first step to address this issue, we need to dissect the 'squeezed middle'.

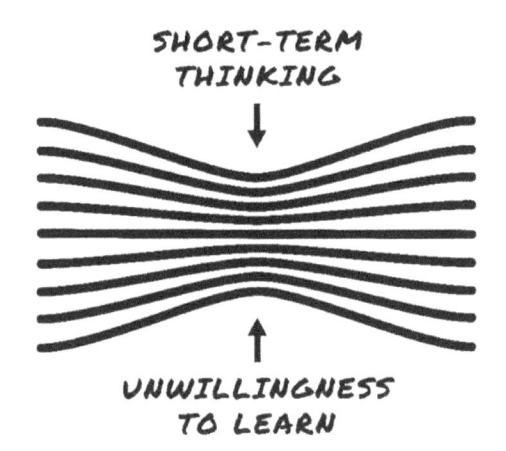

SHORT-TERM THINKING ↓

↑ UNWILLINGNESS TO LEARN

Who is the 'squeezed middle', you may ask? It is you: the pharma leaders who are squeezed from the above by the C-suite who talk about change but don't know how to go about it, and who are squeezed from below by those who directly and indirectly report to you and who are fearful that change means losing their jobs.

If you are a pharma leader reading this and this pain resonates, you're not alone in feeling this way. My team at Cheemia commissioned research to identify the key pain points for pharmaceutical commercial, marketing, and medical managers (I was curious about their pain points and what kept them awake at night), and one of the key findings of the research was that the pharma leaders felt they were 'the squeezed middle'.

Let's first dissect the pressure from above. While pharma's middle managers are eager for change, they are blocked by leadership from the top, who are prioritising short-term results to keep shareholders happy ahead of creating a long-term vision for the business. It is challenging to show an immediate return on investment in change management, and so top leaders resist change because the returns will not be recognised while they are in charge, and they would rather have quick wins that put smiles on shareholders' faces, increase the stock value, and help them get on the next step on their career ladder, than create long-term meaningful change.

This resistance is illustrated through some of the quotes from our research. One such quote was, 'At our company, the owner is a very traditional person. He's way over the retirement age already, and he's like, "Well, why should I invest in things?"'

Creating momentum that motivates senior leadership requires strategic planning. You also need to create drive that meets their short-term needs *and* futureproofs the business.

The research also showed that internal storytelling is an essential skill for 'unlocking' change. One individual said, 'If you go [to the senior leaders] with tech language, digital tools, and dashboards, but you don't sell them the "why", it's not going to happen. Add a little bit of storytelling linked to strategy, and then they will listen.' We will discuss this 'storytelling' component later in this book.

Health needs are increasing and, with that, the demand for medicines is increasing. This means more growth and profitability for pharma companies, and this profitability should not hinder progress and change—

the opposite, in fact—yet pharma is still resistant to change, despite the rapid shift in healthcare decision making, pressure points, and HCP engagement limitations. Senior leaders are not willing to take risks, and although pharma has a great venture capitalist (VC) mindset when it comes to R&D investment and the discovery of new molecules, the sales and marketing functions don't: they are risk averse and unwilling to invest in experimentation and education. Their best attempt at this is an investment in *ad hoc* pilots.

I was talking to a group of European leaders in pharma recently, and they shared that they were sick and tired of pilots. One said, 'We've got pilotitis—an inflammatory condition caused by too many pilots!' They added that pharma is quick to do short-term pilots, and once the pilot is done, nothing happens: these pilots, even if they are successful, don't become standard practice, nor do they get embedded in strategy. Some are tasked with developing long-term goals and strategies, but there is very little connection between short-term isolated pilots and long-term goals. Something gets dropped in between, and this leads to a lack of medium-term strategy development to link the two together.

'Pilotitis' stems from the fact that pharma's commercial and marketing departments are risk averse. There is an inherent fear, conservatism, and lack of belief and vision with regards doing things differently, because they are not sure what the return on that investment will be. Pharma will invest millions of dollars in R&D and new brands that sometimes never see the light of day, but when it comes to the commercial teams and marketing departments, there is no room for error.

This may be to do with the simple fact that at the end of the day, if the team doesn't deliver, their job is on the line. This also accounts for the 'squeezed middle' phenomenon. A cultural shift is therefore needed if pharma is ever going to be open to taking risks. Otherwise, we will just continue to act in line with short-term interests, and there will continue to be a misalignment between departments (i.e., field professionals, pharma leaders, and the C-suite).

Now, let's focus on the pressure from the bottom, which is arguably easier to manage, as the middle managers are directly responsible for how they lead these teams.

Our research brings to vivid clarity these pressures from the field force. According to those pharma leaders, these pressures originate from the fact that field professionals think they know best and are unwilling to learn; to change the recipe they have been using for years. Here is a quote from our research: 'We'll train people in everything connected to multichannel engagement. It will start with a full customer engagement framework. Then, it will go into the more technical pieces: concept gathering, ED data usage, closed-loop marketing, and going on Veeva Engage. But at the receiving end, for the people who are supposed to be the learners consuming this, there was huge resistance.'

You can lead a horse to water, but you can't make it drink, right?

The pressure from the bottom comes in all forms, including, 'I don't want this,' 'This is not the way I work,' 'I'm not going to change,' and, 'I did not sign up for this.'

Our research indicated that instilling the 'why' requires understanding and training, and that those pharma companies that are taking training seriously and investing in change management are reaping the greatest benefits. The crux of the matter is that digital transformation only happens once people have been transformed, and that's the key challenge we're tackling here.

Remote working has had a positive impact on profitability due to reductions in travel time and costs, and to continue this trend, many pharma companies have restructured their field force and reduced it in number. This thinking has unsettled these teams, forcing them to question the longevity of their role and where they fit into digital transformation, if at all. This also means that the members of the field force that remain must be more effective than they were, to plug the gaps. It additionally requires companies to engage with their field force during the restructuring process, specifically with regards technical and soft skills development and mindset and behavioural training. Yet many senior leaders are still using the old, outdated metrics. In 2018 and 2019, the field professionals delivered, say, three to five face-to-face calls per day, and with the shift to virtual and hybrid models, the top leaders expect these call rates to increase, as if the field teams' output and their total working time should be directly proportional to one another. They argue that if the field force is working remotely, travelling is unnecessary, so they just look at the average call time (approximately fifteen minutes), see the field force is averaging two or three calls a day, and become suspicious: what are the field teams doing for the rest of the day?

This way of thinking illustrates a lack of understanding about what's required to engage healthcare professionals in the field. A field professional's task list includes targeting the right HCPs, setting up the calls, researching to understand the networks, and reading policy documents and strategic imperatives to be able to relate to them and talk at their level, amongst other things.

If senior leaders don't trust the field force, if they question the value they bring to the table, and if they speculate over how they spend their time, rest assured that the field force will know about this. Why? Because emotions and beliefs tend to leak out via non-verbal communication and gestures, and the field force is trained to read people and tune into human behaviours. When they pick up on negative emotions from the top, this creates a lack of trust and a rift between the field and the head office—so they 'squeeze the middle' further!

Some pressures from the top are unavoidable and take longer to reduce. If this is the case for you, worry not. In the next few chapters, we will give you some strategies to reduce the pressure from the field force and address their resistance to change. In addition, we provide information to help reduce the pressure from above by sharing information that will encourage senior management to ease their pressure on you and instead support you (or at least leave you alone!) so you can deliver the best outcome and engagement from your field teams.

Also remember to be kind to yourself and avoid putting unrealistic expectations on yourself. The rise in technology and remote working has led to longer working hours and overcommunicating. In these circumstances, it's natural to feel a little drained and overloaded. Life was much simpler twenty to thirty years ago, when the only modes of communication were face-to-face meetings, telephone calls, and the occasional email exchange. Now, everyone is under pressure to catch up on an influx of communications in order to do their job properly. We all work long hours, and the boundaries between home and work have blurred. You could drift back into 'work mode' by checking social media in the evening after the children have gone to bed and start the day by reading emails while eating breakfast. There is no 'off' switch, meaning most of us stay switched on, including at the weekend. This 'never switching off' business will have consequences on our physical and mental wellbeing. We are human beings, after all. We must find an 'off' switch and use it daily.

Checkpoint 2A: Action Steps

- Understand that you are not alone in feeling the 'squeezed middle'.
- Be kind to yourself. Find an 'off' switch and use it daily.

Speaking of unrealistic expectations: pharma marketing mainly uses 'share of voice', 'coverage', and 'frequency of contact' as KPIs, and this results in field professionals creating excessive levels of noise—noise that does not necessarily have a positive impact on HCPs or lead to a change in their prescribing behaviours. Pharma has been lazily stuck on this model for far longer than it should have been, despite the prevalence of much smarter options. This is obvious when we consider the fact that those at the forefront of proper account management and leaders in HCP engagement techniques stopped using 'coverage' and 'frequency' KPIs over two decades ago. The model is pointless because it measures surrogate markers, not performance.

Accordingly, Takeda UK, in a bold move for the market, switched to a Regional Account Director (RAD) model in 2003. At a time when 'share of voice' was the only known approach (and in the face of much scepticism), they decentralised their marketing budgets and decision making to gain faster market access. In a relatively short period of time, the company achieved exponential growth, shifting brands from fifth in market to blockbusters dominating the share in the market.

The RAD model, which gave customer-facing teams autonomy, had a clear strategy: instead of wasting HCPs' time and creating noise, the small, highly effective teams identified and engaged a small number of key decision makers who influenced policy and prescription across a population. Under this model, instead of creating noise, the sales strategy aimed to create mutually beneficial partnerships that led to a win for HCPs, patients, and the company. This was a brave but calculated risk.

Nonetheless, many C-suite executives who are comfortable and familiar (read: complacent) with the traditional approach find it hard to let go of central control. The traditional approach has generated considerable profits for their shareholders, after all, so they are reluctant to take risks or change the status quo, and would instead prefer to assign blame elsewhere (i.e., 'squeeze the middle') if this traditional approach doesn't generate the results they're used to.

This conservatism naturally affects other functions. Traditional marketing strategies are focused on driving brand adoption for all, rather than on healthcare professionals' pain points and narrowing the focus on those who need and want it most, and as a result, their mass promotions and brand focused campaigns appear irrelevant and a waste of time to many HCPs. Clearly, pharma urgently needs to prioritise relevance, demonstrate understanding of healthcare professionals' pain points, and come up with 'no-brainer' propositions that delight the targeted HCPs. When messages are genuinely relevant and delivered at a time when HCPs need them most, marketing teams don't need expensive and noisy 'share of voice' campaigns.

The facts are, healthcare dynamics have shifted significantly in the past twenty years: there are increasing pressures on independent prescribers and primary care practitioners to adhere to formularies, and cost efficiency is high up on health providers' agendas. Yet traditional marketing assumes every account in every country is going to benefit from the brand in question, when the reality is that with these new dynamics, not every account is ready to move forward, and some countries and accounts are going to need and value the brand's offer more than the others. Hence, offers need to be aligned with relevant opportunities and HCPs' needs and preferences.

This was why the RAD model worked so well: it gave the field team the autonomy to make calculated investment decisions and align resource investment to targeted healthcare locations, and in the process, there was no more 'squeezed middle'. Yet many companies still invest equally in every territory and location without consideration of local healthcare providers' priorities and needs. This is a huge waste of resources.

It is also relevant to consider the fact that marketing teams in pharma are grouped by brands who are aiming to communicate their brand to HCPs. Hence, in a particularly large pharmaceutical company, you could have various marketing teams targeting the same HCPs.

In mind of all of the above and in our experience, to grow sales exponentially, you need to:

1. Identify the accounts where there is an alignment between the HCPs' needs and the brand's proposition.
2. Only promote brands that align with the account's needs and priorities.
3. Focus on the 'few who influence many' to gain market access. Communication with these HCPs needs to be hyper-personalised.
4. Have a highly skilled account director work with the marketing teams and medical to co-create personalised, approved communication for these HCPs.
5. Spend time seeking and engaging HCPs on their agenda, not bombarding them with brand messages. Less is more!
6. Invest in the field team to do pull-through only when the formulary and policy is agreed.
7. Use various channels of communication to echo the same message (i.e., the agreed formulary/positive guidance) to a wider

8. Note that pull-through requires higher promotional investment to drive the same message to the wider prescribing base via different channels.
9. Consider tapping into internal channels of communication within the local healthcare economy to create awareness.

To be customer-centric and to stop 'squeezing the middle', we need to evolve and make our marketing strategies account- and customer-focused. We need to wake up and see the world from the customer's perspective first and foremost, not from the brand's perspective.

In pharma, we often talk about being patient centric and say that patients inspire us, yet our actions, decisions, and commercial teams suggest the opposite, in that they are based on profit for the shareholders. The truth is, the organisational structure in pharma is still old fashioned and based on brands, and we need to evolve if we want to be futureproof; if we want to survive. Whatever the course of action we choose is, it is clear that the old system is broken and that we can't just put a Band-Aid on it; it requires a fundamental, creative mindset shift. We need to use the same level of creativity we use in the discovery of new medicines to come up with new models for sales and marketing. We need to learn from other highly regulated industries. Regulations are often used as a reason in pharma for us to not rethink how we do things, but if we look at banking, for example (which is just as highly regulated), it has evolved and improved the customer experience by creating apps and tools that make life easier.

If banking can do it, then pharma is certainly capable of coming up with different models, too.

What's more, if we look at our children and the next generations—Gen Z and the future workforce—they behave and act completely differently to how the earlier generations behaved and acted. Many two-year-olds now have the dexterity to use touchscreens and can swipe on apps before they can speak fluently. In other words, the next generation is already intuitively tech savvy. They have been brought up with tech, and have a completely different mindset towards it than we do. They already have different channel preferences for consuming information. So, we need to change the model to suit the future HCPs. Otherwise, the problems we're seeing (including the pressures felt by pharma leaders from above and below) are only going to worsen as time goes on and the old workforce is replaced by the new generation.

Bottom line: pharma needs to evolve. Our systems (and the way we use these systems) need to evolve. And no, we can't just tweak around the edges; they need to fundamentally change so that they work optimally for the present and future.

To demonstrate the boldness and adaptiveness that is needed here: there is a local family-run business in Edinburgh near the Cheemia office named Mimi's Bakehouse that became a national brand in the UK during lockdown. I met Michelle Phillips, the Founder of Mimi's Bakehouse, and her daughter, Ashley Harley, last year at a business networking event in Edinburgh, where they shared their inspiring story. What I loved most about their approach was their vision and willingness to take risks—and because they took risks, they would do everything needed to bring their vision to life. Michelle and her husband made a bold decision to cash in their pension, sell the family home, and invest the money they needed to in order to establish and expand their business, and when Ashley heard this, she was heartbroken, but also determined enough to move from Dubai to Edinburgh to ensure the investment was profitable and worth it. Their family motto is, 'To do or die.' Their belief, vision, determination, and understanding that there is no going back resulted in them catapulting their business forward. When most hospitality businesses were closing during lockdown, they pivoted quickly and expanded their business online.

What's more, not many small businesses think about branding and brand creation, but Michelle invested heavily in branding because from the outset, her vision was to grow big and be nationally recognised—which meant branding was key. Specifically, Michelle's vision was to create a brand that 'feels like a hug'—and their shop design, packaging, menu, and communications echo that. She recruited staff based on their attitude, not skills, and they have now employed over one hundred and twenty people that they call family. Everything about Mimi's Bakehouse creates a customer experience that feels like a hug—true to Michelle's vision.

Just like Michelle, instead of fearing risk, we need to let risk be our inspiration and do what it takes to make our vision work. We need to work together and evolve as an industry—no finger-pointing or squeezed middles. If we don't, the pharma industry simply won't move forward.

And on that note, allow me to share these paragraphs from Paul Simms, Chief Executive of Impatient Health, often described as the 'pharma provocateur'.

> *Pharma is bureaucratic, conservative, introverted, and unscientific outside R&D. It's stuffed full of people who safely earn above-average wages without the need to risk or invest outside their comfort zone. Despite the slogans and some visible wins in exceptional circumstances (e.g., the COVID-19 response), the majority heavily rely on assumptions and historical processes rather than curiosity, experimentation, and market feedback.*
>
> *This bloated industry will almost certainly trundle along slowly for several years without significant change. The fact that most firms are only now performing structural reorganisations following the COVID pandemic (more than three years after it began) illustrates just how stodgy the internal pace is. In any other sector, nimbler, technology-focused companies would disrupt most customer-facing work with leaner or smarter go-to-market models. Yet despite the fact that pharma's monopoly at the physician interface is not defensible, the enormous cost of entry for new firms to build a portfolio and achieve scale means that margins will stay steady without corresponding value improvement.*
>
> *Quite simply, nobody has to change, and that would be that*

if we were all happy to continue forever with making good profits and relying on our scientific portfolio to keep us buoyant. But if you're reading this, you're probably not happy to just roll over and accept the status quo. You'll be conscious that at the end of the day, we are treating patients, and we have a social and moral duty to do better.

Today, our companies are made of two contrasting halves. One half is driven by exponential science, devising almost magical therapies, now harnessing AI and other tech to improve throughput efficiency in finding new biological medicines, and becoming adept at acquiring external innovation. This half is known as drug discovery and development. The other half is the rest of the company, which is untrusted, slow-moving, and reliant on external, disjointed talent, and is unable to ensure a diversity of patient population, realise iterative development methods, harness creativity or personalisation in communications, or envisage or test alternative business models which integrate consumer technology. The inevitable long-term result is the breakup of many companies—with R&D divisions looking for smarter commercialisation routes and more advanced risk-share partnerships.

Checkpoint 2B: Action Steps

- Think how you can bring a venture capitalist mindset and calculated risk taking to the commercial and marketing departments.
- If your organisational structure is based on individual brands, think about how you can create more collaborative thinking and teamwork between brand teams.
- Consider shifting resources between territories and regions, to invest more where there is greater alignment with healthcare

priorities and growth opportunities.
- Be proactive: turn successful pilots into implementation strategies.

3
Why Isn't the Horse Drinking?

Let's turn our attention to the field teams (end users) and examine their preferences, personas, and work environment before the pandemic.

As we have established, field professionals are generally 'people'-people who enjoy connecting and communicating with other humans. They love variety in their day, and the idea of travelling to different locations and meeting new people energises them. Field professionals initially join the industry with a science degree or a scientific background, or with experience as a healthcare professional, and they opt for a field position because they enjoy communicating with and selling scientific content to healthcare professionals. They enjoy variety, lack of routine, travel, varying locations, and interaction with many people. Traditionally (before COVID, that is), the main mode of communication in their role was face-to-face meetings, and they had ongoing coaching and training for this from their first-line managers. Over time, they became masters of this face-to-face communication.

Field professionals generally do not gravitate towards Excel sheets. They don't like working in an office, having the same routine day in and day out, or admin. If they loved technology, spreadsheets, and analytical tools, they would not have opted for a sales career and would have instead pivoted to a career in a head office function, like marketing.

Of course, these are sweeping statements, and not all field professionals have these characteristics—some do enjoy admin and have advanced analytical skills—but generally speaking, the majority are more comfortable with talking and interacting than they are with using technological tools and platforms. Hence, asking the field force to adopt new advanced technology was forcing them out of their comfort zone. Before this, they were used to walking into a room and using their personality and charisma to connect with individuals. Trying to do that through remote channels was not only alien to them, but it also caused them to worry about their performance. This made their interactions more challenging, and, sure enough, they could not recreate the same chemistry via their new tools that they used to create when they saw HCPs face-to-face.

Let's use the analogy of driving a car to illustrate this point. Remember your first driving lesson? It was a really scary experience, right? You needed to learn how to control the car, use clutch control, and do the 'mirror, signal, manoeuvre' routine. Most of us questioned if we'd ever manage to get our heads around it all, yet when we look back on this after confidently driving for years, it seems a distant memory. Now, we can drive with ease and get to our destination without even thinking about it. But that doesn't negate the fact that, in the beginning, our experience and confidence were a far cry from that.

In the same way, to ask field professionals to shift from interacting with customers face-to-face to interacting with them remotely, *and* to use new digital tools and channels, is like taking them out of their company car, putting them in a Formula One car, saying, 'Off you go,' and expecting them to speed off into the distance without a hitch. To drive a Formula One car, you need special training and practice, and it takes time to get your head around driving it competently. If you've not been trained to drive a Formula One car specifically, you'll probably stall it or end up spinning off the road, regardless of how competent you are in driving a 'normal' vehicle.

So, like the journey we all went on when we first learned to drive a car, getting field professionals to adjust to this new way of operating is going to take time. They're going to make mistakes and they'll probably need an instructor sitting next to them coaching them through the process, and through practical experience (i.e., through them using the new tools), they're ultimately going to wrap their heads around it—but this isn't going to happen overnight. Yet some people expect pharmaceutical field professionals to be able to pick up these digital tools, absorb the content by osmosis, and run with them within a matter of weeks or months, without an initiation period or a roadmap.

This was, and is, unrealistic. If you'd just given them the Formula One car without a coach or instructor, the car would be destined to sit on the tarmac unused (especially if you gave them no practical reason for them to try to figure it out themselves).

They Are Unmotivated

Let's now turn our attention to field teams' motivation for using technological tools.

I use technology in my work every day, and I am quite an expert in using technology to engage with HCPs. However, I find myself struggling with the remote for our TV, much to the amusement of my husband. Why? It's not because I lack the capability. I simply have no interest in learning how to use it. I'm not particularly invested in TV, the remote control, and all the various channels it offers; I prefer to watch streaming networks, and I know how to access them.

In general, people tend to put effort into learning things when they are genuinely interested in and can personally see the benefit of doing so. This is never more evident than with digital transformation in pharma. Motivating teams to tackle omnichannel requires effective leadership and an understanding of individual motivations. I am sure if you reflect, you will find at least one thing (either in your personal life or your professional life) that you are not willing to master because you simply don't see the need to.

Field teams are no different to you and me: they need to have a big enough reason to change, and they need to see the benefit of changing, in order to be motivated to do so.

They Are Afraid

So, what are the field teams' pain points? Why is the horse who has been led to water not drinking?

Despite massive investment, field force tech tool and digital content adoption is very low across the industry. The field force is generally feeling uncomfortable with and overwhelmed by digital tools, but they are not sharing these feelings with their superiors because the expectation is that they need to deliver, plain and simple. Essentially, they don't want to draw attention to their Achilles' heel. This means that senior leaders are often blissfully unaware of these pain points and how they impact field performance.

Allow me to share a personal story to demonstrate my point here:

In September 1986, I began my nurse training at Maidstone Hospital. After six weeks of initial training, I started my practical training on the John Day orthopaedic surgical ward, where I would be working for the next two months. By this point, I had been living in the UK for just over a year, and was finding it easier to hold conversations in English, although I wasn't yet proficient.

When learning a new language, you often understand more than you can express. For me, reading written information was easier than understanding spoken words. I believe this was because reading allowed me to take my time, while oral conversations required quick responses but more complicated processing. Over time, I improved in reading people's body language, but I struggled with phone conversations, especially if the person had a strong accent.

The ward had twenty-eight patients and, as student nurses, we were assigned to the care of approximately seven patients during each shift. At the beginning of the shift, we received a detailed handover to learn about all the patients and their care needs. Remembering all the names (which were foreign to me) and details (surgical procedures I had not heard of before) was challenging for me, so I would write everything down and place all my focus on the seven patients under my care.

The clinical aspect of nursing came naturally to me and was enjoyable. However, answering the phone posed the biggest challenge. I felt nervous about not knowing all the patients' names and their conditions, fearing that I wouldn't understand the person on the other end and embarrass myself. So, whenever the phone rang, I would distract myself with other tasks, hoping someone else would answer it.

Despite my efforts to slip under the radar, the charge nurse, Eddie, was sharp, and noticed my avoidance of phone calls. After working with me for two shifts, he called me into his office and asked why I wasn't answering the phone. I explained that English was my second language and I felt anxious about understanding callers. So, Eddie made a decision: 'From now on, every time you're on duty and the phone rings, I want you to answer it.'

I gulped, terrified. Sure enough, it was a frightening experience: Eddie had really pushed me out of my comfort zone with this one simple instruction. However, I managed to overcome my fear, and I gained confidence in answering the phone without being scared after a while.

Eddie recognised the fact that effective nurses needed not only clinical knowledge but also strong communication skills in order to handle different situations and stakeholders. He also understood my fear—he did not deny or invalidate it—and he created a safe space for me to practice, make mistakes, and learn. As a first-year student, it was okay to make mistakes, and I had the support of staff nurses if I ever got stuck.

I share this story because my fear of using the phone prevented me from utilising this (very helpful) tool, but Eddie's dedication to pushing me out of my comfort zone and the support he provided helped me to embrace it quickly. Similarly, pharmaceutical companies have provided their field teams with the digital tools they need in order to engage remotely, but they haven't provided them with a clear roadmap on how to use them, nor are they supporting and guiding them through that transformation. Instead, the field teams are expected to 'just make it happen', but they don't know where to start—so, in absence of a clear roadmap, they revert to what they know: face-to-face interactions.

Field teams feel too vulnerable to speak up about their struggles, since their livelihoods depend on their performance. Drawing attention to their capability gaps would question the necessity of their place in the workforce, so for them, staying quiet and below the radar is easier than risking their job security, even if they run the very real risk of consistently falling short of their superiors' expectations as a result.

The leadership, marketing, and medical departments are very excited about these shiny new tools and how they will transform HCP engagement, but these very tools have left the field teams feeling vulnerable. They wonder, *Are these tools going to replace me? Where do I fit into all of this? Am I going to lose my job? Is the future of communication with healthcare professionals going to become completely digital?* These fears create resistance to change, and these fears are all rooted in the perceived possibility that soon, there will be no need for human-to-human interaction at all. These fears are unfounded, but understandable.

More relevantly, the pain and fear the field teams are experiencing is real to them, so don't deny it, brush it under the carpet, or sugarcoat it. It is not their fault they feel this way; it is their circumstances. So, make time and seek to understand it and connect with them on a personal level.

They Are Demoralised

In 2022, in collaboration with Sebastien Noel, Director of Multimedia Strategy at Veeva, we invited digital and commercial leaders from global pharmaceutical companies to discuss the challenges facing the industry with regards to HCP engagement in a post-pandemic digital environment. We wanted to hear the experiences and insights of the companies who were grappling with the adoption of digital technology and omnichannel engagement from the human perspective of the field teams.

Here is a brief extract from this report that we published, in which Sebastien explains Veeva's research:

> *There is a change in the channel mix that we see with healthcare professionals today, for your field teams. When I say 'field teams', I speak mainly about commercial but also medical, because we know that the MSL teams are also more and more important for you. How you include them in the whole engagement strategy makes more and more sense. The channel mix before COVID was ninety to ninety-five percent face-to-face interactions, and you of course had some emails. You had a few remote meetings for some companies as well, but very, very few pharma companies worldwide were really thinking seriously about this whole hybrid approach with face-to-face and digital.*

GLOBAL CHANNEL MIX

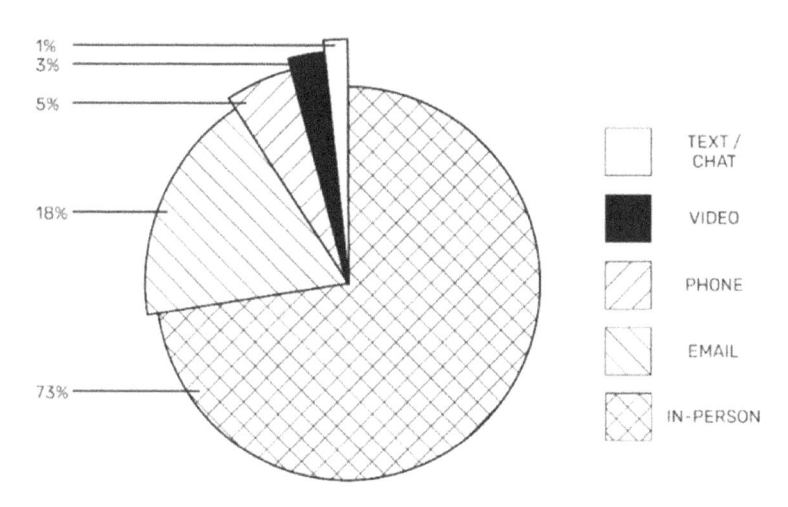

- 1% TEXT / CHAT
- 3% VIDEO
- 5% PHONE
- 18% EMAIL
- 73% IN-PERSON

This is data from over one hundred and thirty million quarterly

HCP-field interactions across eighty percent of global biotech and pharma companies, so we can really understand what's happening with all the pharma companies today and the interactions and channels being used by the field teams [to communicate] with healthcare professionals. When you look at the global channel mix, the key message is that the world is about seventy percent face-to-face and thirty percent digital. And when you look at digital, it's a lot of emails and some phone calls. We also have video calls—Veeva Engage, Zoom, and Teams—and you also see text and chat messages emerging.

I believe this mix has already evolved. We may have a bit of over-index with face-to-face right now, but it may decrease in the future; we will see. And we know that healthcare professionals are really eager to have the right mix. Even if they want to keep face-to-face, they also want more digital. For them, it makes sense. They want the best of both worlds. So, that's something we need to take into account.

UK CHANNEL MIX COMPARED TO OTHER MARKETS

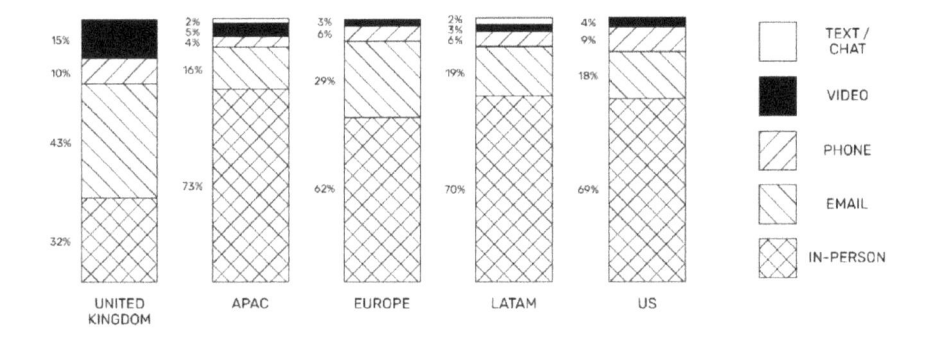

	UNITED KINGDOM	APAC	EUROPE	LATAM	US
TEXT / CHAT	—	2%	3%	2%	4%
VIDEO	15%	5%	6%	3%	9%
PHONE	10%	4%	—	6%	—
EMAIL	43%	16%	29%	19%	18%
IN-PERSON	32%	73%	62%	70%	69%

And when you look at the breakdown by region, you see that country-by-country and region-by-region, it's already quite different. I want to take the example of the UK, because it's maybe the country that is showing us what the future could look like. You see that face-to-face is already way lower; it's about thirty-two percent. You have a lot of emails, you have way more video calls, and you have more phone calls, as well. So, this is going to evolve. I think every month, it's going to change.

Yes, we know the world is changing—you're working in your respective organisations on the strategy to go multichannel, and from multichannel to omnichannel, and making sure that you have a better customer experience for the healthcare professionals—but what about your people? What about your field teams, their managers, and also the other people in the marketing teams, who are working on content? Who is making sure that the whole organisation is going through this change?

The challenge is that barriers to conducting face-to-face interactions have arisen, not just because HCPs are under unprecedented time pressures, but because HCPs' communication preferences have changed. They now expect to communicate via a mix of channels, including face-to-face. Hence, using only face-to-face interaction as a channel of communication will limit access to engagement opportunities for field teams.

In some countries, the field team feels added pressure in this respect because the rate of rejection is much higher now than it was before, and this rejection, if not channelled and addressed effectively, can impact their self-esteem and mental wellbeing. Basically, it is really, really demoralising, especially without that all-important support network in place.

If you are interested in reading this full report, please head to www.theomniadvantage.info/report or scan the QR code on the following page.

The Content Is Not Relevant

Another factor that exacerbates this problem is the fact that the digital content developed by marketing is not always fit for purpose. This usually happens because the content is focused on the brand's features (rather than on the HCP's pain point) or is not geared to communicate a compelling story. This means that the field teams either don't like the content or, after observing HCPs' reactions to and lack of interest in the content, they take the initiative and stop using it.

When I interview field teams, they openly admit they don't use digital content. They know using visual data would improve communication and recall, but the digital content available is often not relevant.

The problem is that many senior leaders don't consider the 'user experience' of the content that is being created. I am sure if they knew, they would be horrified, but they remain unaware because most of the content that is developed for field teams is not examined at the end user level for validity or experience.

Let me give you an example: a pharmaceutical company once engaged Cheemia's services because they wanted their field professionals to increase their monthly uptake of e-permissions as a KPI. They were frustrated that their field teams were not achieving this. Upon taking a closer look, the problem became clear: the process for gaining these e-permissions was complex and disjointed. The automated system that had been designed to capture these e-permissions via the CRM system was not working, and had not been for some time. The alternative process to plug the gap while they were waiting for the CRM to be fixed was too complex and time consuming to implement. Plus, the field force did not have any approved digital content to give to the HCPs, so the hook that allowed them to ask for e-permissions was also missing. As a result, the field professionals focused on selling rather than capturing e-permissions.

Although on the face of it the field team were responsible, the root cause of the problem was completely separate from their role.

My point is, as with manufacturing processes, marketing teams need to test their outputs along the production line and check further down the conveyer belt to ascertain whether they are working and being adopted. If they are not, they need to check to see what is causing the hiccup. In manufacturing, the standard practice is continuous improvement: they test the production line every step of the way and look for the simplification and improvement of their processes. After the product is developed, they test it before putting it out on the market. When it comes to pharma marketing materials development, however, there is detailed testing and checks for compliance, but there isn't this kind of rigorous testing to check the field team's user experience, nor do they check the steps for the field teams to access and retrieve the digital content.

If the marketing team tested the internal steps with the end users, they would pick up on these challenges and be able to increase the uptake of the materials by simplifying the process, so the end user could easily access and use the content.

The Tools Are Impractical

Our research also showed that these field teams found corporate technology platforms, such as CRM systems, to be clunky and user-unfriendly. Yet at the same time, they said they loved business-to-consumer (B2C) digital tools, such as Netflix, Amazon, Facebook, and Uber, and that they used them regularly for ordering and consuming digital content and communications in their free time.

The platforms used in pharma are often complex and time consuming. The thought of using some of these platforms is enough to bring on a sudden headache for the end user. Plus, the field teams' early experiences with these platforms will also affect their judgement.

I can personally relate to this: I remember when the first version of Salesforce CRM was rolled out in 2015, it was not as user friendly as it is today. I once had a urology meeting booked in Aberdeen, and it took me less time to do a five-hour car journey from Edinburgh to Aberdeen to conduct the meeting than it took me to record the interaction on the CRM system. Several calls to the helpdesk, going back and forth to validate HCPs with the Binley's database, were required before the calls could be recorded. I was so frustrated that I wanted to throw my laptop into the nearest river, and I was not alone in feeling this way.

The digital tools that are developed for field teams are often not designed with the end users' preferences in mind. Instead, they tend to be designed based on leadership reporting requirements and dashboards, and so they fail to create a positive end user experience. UX design principles are rarely applied to developing in-house tools for field teams.

The question here is, is the CRM system being used to meet the needs of senior leaders? Is it being used to measure irrelevant and outdated KPIs? Or is the CRM system being used to track customer activity, gather insights, and help the end user to play their part in orchestrating a positive customer experience?

Much of the valuable insight gathered by the field force daily is not captured on the CRM system to build customer insight. Why? Because the field force sees the CRM as a monitoring tool—as the Big Brother who is watching them—and not many companies have explained the value of the CRM system in building our information repository.

To increase the field team's adoption of corporate technology platforms, the digital tools in question must be designed to create a pleasant user experience, like the B2C digital tools the field teams are already consuming. We need to design digital tools that delight the end users. Then (and only then), the end users will organically be more willing to adopt them. If we make pharma tech platforms more user friendly, evaluate the end users' experience along the journey, and make them as easy to use as Amazon, Netflix, or Uber, the field teams will embrace these platforms, not avoid them like the plague.

They Are Given Little Autonomy

Pharma companies employ bright individuals with scientific backgrounds who are naturally good communicators, problem solvers, and fast learners. Yet corporate communications are generally done via one-directional leadership: the company employs the field team and tells them who to see (targeting), how often to see them (frequency), and what to say (key messages), and the field team's role is to go and communicate this to the healthcare professionals. In other words, they are expected to just 'get on with it' and deliver outcomes. This one-directional leadership creates a parent-child ego state relationship between the company and the field team (despite the field team being more than capable of independent decision making and adult-to-adult interactions) and, if continued, it leads to the deskilling of the field team and overreliance on direction from leadership. Thus, to unleash the field team's potential, the field team needs to have a greater degree of autonomy in decision making, targeting, and creating feedback loops to share HCPs' insights with marketing and head office.

Osman Dağgezen, author of *Omnichannel Customer Engagement in Pharma*, used a great example to illustrate this point on a post he shared on LinkedIn in 2023. He said:

> *Ever lost your way by a GPS while driving? I certainly have. So focused on the voice instructions, I sometimes miss what's right in front of me—a clear, simple road sign pointing the way.*
>
> *This experience always reminds me of the dangers of*

micromanagement in our professional lives.

Just like a GPS, micromanagement can make us dependent, stifling our ability to think and act independently. When instructions are overly detailed or unclear, creativity, innovation, and growth can suffer.

In our journey to success, let's not forget these 'traffic signs'—the ability to think, adapt, and make decisions on our own. Embrace autonomy, foster trust, and remember that sometimes, the best route isn't the one dictated by a GPS.

When I worked as a senior leader in Pfizer, we discussed challenges as a team and made collective decisions. This was a positive and inclusive corporate culture, but it got me into the habit of consultative decision making. When I moved to Takeda as a Regional Account Director, on the other hand, the culture was more entrepreneurial: I had control of the budget and was responsible for P&L, and reported directly to the Board. We did not have monthly meetings, and we worked independently.

One day, I approached one of the senior leaders to seek his input relating to a business-related decision, and to my surprise, he challenged me as to why I needed his input. He argued that they paid me handsomely to make these types of decisions (good or bad) and live with the consequences.

That was a turning point for me. The autonomy and trust he put in me (and the responsibility that went with it) unleashed my potential as a leader. It allowed me to experiment and explore new ideas. I started making decisions, and as I did so more and more, my confidence grew exponentially. If I made a mistake, I owned up to it and looked for the lesson in that experience.

Six months later, I was sitting with my peers in the Heathrow airport lounge while waiting for our Edinburgh flight. It was there that I ran into Stan, my ex-manager at Pfizer. He approached me tentatively and enquired, 'Is that you, Mehrnaz?' He said he did not recognise me: I had my hair down,

I was laughing with my colleagues, and I looked relaxed and confident. This was the result of me being given more autonomy and trust in my workplace.

Instead of micromanaging and creating parent-child ego states, companies will gain an advantage by treating the field team like adults—i.e., giving them the autonomy to shape HCP communications, seeking their input in targeting, utilising their field insights, and welcoming their input for content co-creation.

Change Is Being Dictated to Them

Many pharma company leaders assume change is orchestrated from the top down. They communicated to the field team that they would need to adapt to the technology (and expected it to happen) and thought it was job done. However, change does not happen from the top down; change happens at the individual level. It is like planting a seed: we can encourage growth by creating fertile soil, watering the seed, and putting the seed in a sunny position, but at the end of the day, the seed needs to grow from within.

The same applies to humans and the field team. Growth starts within, and if field teams are not adapting and changing, there is something internal that is preventing that adaptation and change from happening. Those internal causes need to be understood and addressed before any further action is taken.

I recall talking to Koen Janssens, Global Associate Director of Field Force Effectiveness, about the effectiveness of his field force in the summer of 2020. Koen said he had invested in various training initiatives, and every time he did so, he would ask the field team, 'Are you clear about what you need to do with this?' The field team would respond, 'No, we need to go away and figure it out.' As it turned out, the training concepts were too theoretical and lacked practical relevance, and were therefore difficult to apply to the reality the field teams were facing. This left the field teams more confused and unclear than ever, because the training had failed to provide them with a clear roadmap.

The training and solutions offered to the field team need to resonate with the reality they face in the field; they need to be clear and practical in nature.

They Are Not Kept Up To Date With Communications

Marketing needs to consider the implications of using email communications on behalf of field teams and how this could break down trust. I was talking with a senior KAM a while ago, and he told me that healthcare professionals were receiving repeated emails from 'him'. He had no idea that these emails were being sent out on his behalf by marketing, and the healthcare professionals were contacting him saying, 'Is everything okay? These emails don't even sound like you.'

Sending lots of email communication with approved content may seem like a good idea, but it can have a negative impact on healthcare professionals: it breaks trust when they know the individual would not have written the email in that style, and an email that is so geared towards marketing and is not personalised could damage the hard-earned relationship between the field professional and the HCP.

They Are Given Meaningless KPIs

The field team's behaviours can be shaped by KPIs, since what you measure is what gets delivered—so be mindful of what you measure!

Measuring surrogate markers could encourage ineffective behaviours. For example, some companies are measuring channel choice as a marker of digital adoption and are giving higher scores for, say, a digital video interaction via Teams than a phone call, but to be truly customer focused, we need to give HCPs the freedom to choose the channel that suits them best. Thus, I recommend not predetermining the channels or being too prescriptive about how the content must be communicated. Measuring the customer's experience or the quality of the relationship is much more meaningful than measuring interactions by something as arbitrary as channel type.

Remember, it isn't about the tools themselves; it's about what we want to achieve *using* the tools.

Let's look at an example of building an IKEA bookshelf. The result we want is a bookshelf we can put our books on. There's no point in measuring how many times we used the screwdriver or the Allen key; the question is whether the bookshelf is built and fit for purpose.

James Harper, Founder and Managing Director at twentyeightb, has a valuable perspective on this matter:

It has only been over the last few years that I have really come to appreciate and now live by the digital transformation mantra of 'mindset, skillset, toolset'.

I am a toolset guy. I love technology. In fact, I set up twentyeightb back in 2010 because I was an early adopter of the iPhone 3 (launched in the UK in 2008) and I was seeing the app marketplace explode with creative solutions, and I saw opportunity for pharma in all of that. I already had the mindset, and I simply forced myself through the skillset because my new business needed me to.

But not everyone is the same, and if we keep jumping to the toolset without addressing mindset and skillset, then not only will we leave some people behind, but we will also find that the toolset has been left on the shelf with no one using it.

The data bears this out: the Veeva Pulse data from 2022[1] shows that seventy-seven percent of content generated for field teams is rarely or never used. In contrast, the same Pulse data shows a 2.5-times increase in new patient starts when digital content is used. The effectiveness of digital content is also borne out by the STEM healthcare meta-analysis[2] that looked at iPad usage from 2011 to 2017 across every EU and North American client. The data shows that compared to a flat-PDF-style sales aid, enabling a field team with a fully interactive one, where both the rep and the HCP interface with the content, results in a seventy-one percent improvement in the frequency of a good sell outcome (which is when the customer agrees to behavioural change or, in simpler terms, they say 'yes').

So, how do we close the gap between current usage and desired usage?

For me, it is all about winning hearts and minds and ensuring competence and confidence in using the content and technology—so mindset and skillset first. Why? It is simple: if I don't believe in the value of something, then why on earth should I use it? And if I cannot use it easily as part of my daily workflow, then why would I use it?

I am not going to address skillset, as I rely on experts and strategic partners like Mehrnaz and her team when it comes to upskilling field teams, and this is not the place to get into detail

---

[1] For the Veeva Pulse Field Trends Report from April 2023, head to https://bit.ly/field-trends-report.
[2] The meta-analysis to which James Harper refers is the STEM Healthcare Meta-Analysis of Observed Calls Across North America and the EU (2011–2017).

on toolset. So, let's talk mindset.

When it comes to our HCP customers, as an industry, we are getting better at understanding and addressing the 'job to be done'—that is, what are we really helping the HCP to do. But when it comes to our internal customers, it feels like we are a million miles away from that thinking.

So, what is the 'job to be done' when it comes to field teams? Well, frankly, it is the same as it is for everyone: make my life easier and help me do my job better so I can earn more money, spend more time with my family, be in the top ten, and get that promotion. It is our job as content providers, technologists, and data wranglers to demonstrate how our solutions can do this for field teams.

So, how do we do it? How do we win over the hearts and minds of the field team?

1. Take them on the journey, co-create the tools and content you want them to use with them, and create champions and stakeholders in the field, just as we do by bringing HCP KOLS onto advisory panels. Sure, we want them to provide feedback on how we can best use the new data to promote our brand, but we also want them excited and looking forward to sharing the inside knowledge with their peers and colleagues.

2. Make sure whatever you are producing or asking the field team to use fits into their daily workflow and how they engage with their customers. I swear, if I see another sales aid that has been produced without alignment with the sales process that the reps are trained in and asked to implement every day, I will scream. If the agency creating your sales aid has not asked for a copy of the sales training deck and asked how long, on average, your field team gets in front of the customer, you need a new agency.

3. Think about how your content can help the field team do

their job, like a sales aid that can type for their customers, capture insights, or autocomplete the call notes field for them. We all love things that make our lives easier.
4. *Evaluate and share successes, however small. You could also share failures as learning opportunities, but to do so effectively, you need a culture of shared responsibility free of blame.*

Here is an example from way back in the early days of CLM in CRM. The presenter before me at a pharma digital conference talked about how sharing data in a non-judgemental way had significantly enhanced their sales figures.

The northern field team in their country had historically always outsold the southern team. The southern team had claimed differences in access and travel time—you know the usual excuses by now. And perhaps there indeed was a real reason for the difference. Yet after they adopted Veeva CRM, the marketing team identified a trend in the use of their tagged content (where each slide of the sales aid is associated with a brand key message). The northern team were talking efficacy first and then safety second, whereas the southern team was talking safety first and efficacy second. All the marketing team did was share this insight with the southern team and suggest that, as we all know, you need to convince someone that something works before you bang on about how safe it is.

Over the following year, the data showed the southern team had started talking efficacy first and safety second, and over time, the sales gap closed.

Getting this right with the field team is so important. COVID has highlighted just how stretched a lot of healthcare services are, and has resulted in long waiting lists for patients, so it is now a moral obligation for us, as an industry, to ensure we do not waste the time of our HCP customers. So, we need to *ensure the field teams are empowered with great content and with data and insights on their HCPs that enables them to make informed decisions and be the orchestrators of great customer experience.*

For companies to get the maximum return on their tech and digital investments, their field teams need to be confident in using the tech to recreate the same chemistry they have with HCPs online as they do in face-to-face interactions. Technology becomes an enabler for this if the professional in question feels confident, tech savvy, data literate, and able to engage and communicate with HCPs using multiple channels. On the flipside, if they are unfamiliar with the digital tools and feel overwhelmed and uncomfortable when engaging with them, they will revert to what they know, given the option: face-to-face interactions.

In summary, the horse isn't drinking because the horse is feeling uneasy, overwhelmed, and uncomfortable over the resource it's being presented with. So, companies must take the time to really understand the field teams' pain points, act on their insights and feedback, and invest in the right skills development to help them embrace digital transformation. Companies that make this investment will see a massive improvement in their field teams' capability in a very short period of time, that's for sure.

Checkpoint 3: Action Steps

- Consider the field teams' personas, and design the tech tools based on their preferences.
- Understand the field teams' motivations and why they choose not to do something. Focus on the benefits that this change brings to them personally, and help them do their job as they navigate this transition.
- If you encounter reluctance amongst your field teams regarding

using digital channels, take the time to understand their reservations and fears and acknowledge their pain.
- Seek to understand the shift in channel mix from the field team's perspective, as well as how rejection affects your field team's mental health and wellbeing.
- Check that your content is relevant to HCPs and pay attention to field user experience. Listen to their feedback and simplify the digital content usage processes.
- Ensure the digital tools, the content created for the field teams, and the CRMs are user friendly and easy to access, and they fit into the field team's normal workflow. They need to delight the end user and be designed to make their job easier.
- Explain why we collect the data on the CRM and what role the field team can play in that journey.
- Create a balance between providing autonomy and offering support that encourages innovation and feedback. Instill an adult-adult relationship between field teams and office functions.
- Change happens at an individual level, so make training practical and relevant to the reality the field teams are facing.
- Inform the field teams if you plan to send email communications to HCPs on their behalf. Overusing emails without consent from the field team will damage their relationship with HCPs and lead to the withdrawal of e-permissions.

4
HCPs Are Human, Too!

Let's take some time to think about HCPs, how their world has shifted, and what their expectations are now regarding engagement and communication with the pharma industry.

As we have previously touched on, the pressure on HCPs' time has increased because their workload has skyrocketed. They face increased demand for services and long waiting times while they simultaneously juggle crippling staff shortages. As if that wasn't enough, they also have an increasing administrative burden to add to the pile. Hence, HCPs need to protect their time now more than ever before.

Since our industry and the ways in which we communicate have gone through major transformation in recent years, healthcare professionals' working environments and preferences have seen a seismic shift accordingly. Hence, to connect meaningfully with HCPs, we must not only examine their new channel preferences and information consumption behaviours, but, more importantly, we must understand their pain points, needs, and availability. To do this effectively, we must first pay attention to what they are telling us before we rush into planning.

Recent reports by DT Consultancy,[3] Reuters,[4] and IQVIA[5] show a fundamental shift in healthcare professionals' channel preferences due to a combination of factors. For one, HCPs have started using technology for patient engagement and remote consultations, and so are now more familiar with this communication method. In addition, there is a new workforce in healthcare: Gen Z (as we have established) is, broadly speaking, much more tech savvy than previous generations, and so the new workforce is more open to digital channels for engagement and prefer a mix of face-to-face and remote engagement channels (though they lean towards digital channels because they can access information easier and quicker).

When HCPs engage with pharma on a topic of interest, they have questions. This is a good thing, and it can create a further five or six digital touchpoints. For example, following an engaging video meeting, HCPs may have a medical information request for additional data, or they may request that copies of any digital assets that were shared during the meeting are emailed to them. If they plan to share the content with other stakeholders, they might ask for additional digital assets, make an email introduction between the field team and other stakeholders, or request a follow-up meeting or presentation at a group meeting. These increased digital touchpoints with HCPs lead to faster decision making and adoption, so they are certainly a good thing, but they also mean that field professionals need to be very adept at completing these requests quickly and efficiently.

When a HCP receives a request for a call or a meeting, they stop and ask, 'Is this going to help me?' because they only sometimes get value from their interactions with pharma. Historically, interactions between pharma and HCPs have been based on pharma's agenda rather than the HCPs', and the average customer experience score across the industry confirms this.

The 2022 DT Consulting report on the current state of customer

---

[3] *DT Consultancy—The State Of Customer Experience In The Pharmaceutical Industry*, 2022: HCP Interactions (Feb 2023).
[4] *Reuters Events Pharma: Annual Industry Trends Report 2023*. A new holistic and purpose-driven approach should achieve better and more equitable outcomes. Thinking bigger in 2023 to drive better outcomes.
[5] IQVIA white paper, *From Surviving to Thriving: Changing the Paradigm of HCP Interactions* (Feb 2023).

experience in the global pharmaceutical industry[6] analysed individual companies, and there were some interesting takeaways:

- Overall, customer experience (CX) is just 'average'. Globally, pharmaceutical companies are failing to give HCPs an 'excellent' experience. None of the pharma companies in the study scored 'excellent', and not one is setting the pace. That means there is lots of room for improvement.
- According to this study, 'human use of technology provides the best CX'. HCPs preferred direct online video conferencing with a medical rep, and found that purely digital channels just don't pull their weight.

My favourite excerpt from the report is this: 'To ensure they're protecting the right experience, firms should implement a continuation of rep training and development that focuses on enabling reps to embrace digital capability and understand the continually evolving customer engagement models.'

I couldn't put it any better myself.

The report recommended that firms start anticipating the future of customer engagement. Pharma must create more personalised campaigns and double down on segmentation and targeting, and we also need a solid foundation of data to better understand HCPs' changing preferences. HCPs want and need the pharma industry to wake up and work on the human element of digital transformation, so pharma customer experience needs to gear up for a hybrid world and not only be mindful of not wasting the HCPs' time, but also be sharper at grabbing their attention. Field teams have seven seconds to grab HCPs' interest when using remote channels, and if they fail to do so, HCPs will end the call much quicker than they will end a face-to-face interaction.

As we have established, pharma field teams also need to understand the healthcare professionals' environment, pain points, and data needs, as well as the unique brand's benefits, propositions, or information that can

---

[6]DT Consultancy—*The State Of Customer Experience In The Pharmaceutical Industry*, 2022: HCP Interactions (Feb 2023).

directly address those pain points. They need to stop and think about what they've specifically got to deliver that merits the HCPs interrupting their patient consultations or staying a bit longer to talk to them. If they cannot think of something that they offer that will do this, there's a problem. Being relevant is fundamental. If you are not relevant, fighting for a digital share of voice is futile. HCPs are overwhelmed by digital data and communications and don't have the time or capacity to digest all the information that comes their way. For this reason, they welcome input from pharma professionals who they trust, understand their challenges, and have something of value to offer.

Often, the underlying cause of HCPs' low engagement (and field professionals' reduced access to them) is the pharma content they are presented with not being relevant. Some companies are considering replacing the field force with MSLs to overcome this access issue, as they assume that MSLs are a better resource for delivering content to healthcare professionals, but the value MSLs bring to HCP interactions is contingent on each situation, where the brand is in its lifecycle, and the complexity of the scientific information required to communicate to HCPs. MSLs are good at communicating complex scientific data and they are analytical, but they are not skilled in interpersonal skills, selling, or account management, and, similar to the sales team, they are reluctant to adopt digital channels and content. In other words, the problem is not with the field force. The bottom line is, when pharma's messages are relevant and delivered by field teams succinctly, they land with impact and engage HCPs.

Let me share a comment left by Thansi (Athan) Tachtatzis, a senior pharmacist in medicine management at the NHS, on a post I recently shared on LinkedIn about the role of the pharma salesforce. Athan said:

> *A true sales interaction occurs between two people or groups. It couldn't be truer than in healthcare. Once upon a time, we used fancy information sheets and prospectuses. Times have changed, so we use digital tech. But at the heart of everything is*

a person, as you rightfully say. Enable sales teams with appropriate digital awareness/skills.

Over the past three years, we have seen that digital-only launches tend to fail because they don't address the human element of connection, relationship, and account management.[7] People buy from people, and their buying decisions are influenced mainly by emotion.

For example, many of us use iPhones. Technically, the iPhone does not have the best camera or operating system, yet many of us still opt for an iPhone again and again. I do amateur photography using my phone, and for that reason, Samsung would have been a better choice for me, but I stick with iPhone because I'm making an emotional decision. I am familiar with the operating system and I have an emotional connection with the brand, and no amount of advertising or information will change that.

HCPs are the same. They make emotional decisions based on their relationships with people and companies. They get bombarded with advertising and digital content left, right, and centre, so they find that talking to people who they connect with on a personal level helps them to make informed decisions. It is not about the sheer volume of content; it is about the relevance of the content and the emotional connection with the person delivering it (who normally knows when and how to deliver it).

So, to summarise: understanding healthcare professionals' pain points is essential if pharma wishes for their messages to remain relevant. We must figure out the key things that matter to HCPs and then communicate the messages that resonate with those pain points.

Allow me to illustrate this with an example:

We have three key messages for our brand: one is about efficacy, one is about tolerability, and another is about the cost. When talking to different customers, we need to lead with the message that resonates with them the most. In other words, the field force needs to have the flexibility to use the

---

[7]Iskowitz, M. (2022): *Built to fail: Pharma's digital product launches plagued by single-digit success rates.* Published in *Medical Marketing and Media.*

brand messages in the order that is most relevant to each healthcare professional at the point in the decision-making process, because if we address their pain point immediately, this will create more meaningful interactions. When you hit the sweet spot and work along the same agenda, suddenly, HCPs open up and are fully engaged, and often, they leave the meeting with ideas for actions to drive the project forward.

Clearly, we must move away from the approach of dumping marketing messages on healthcare professionals in the hopes that something sticks. Instead, we must deliver a valuable message addressing the given healthcare professional's pain point. Specifically, there are three elements required for effective HCP engagement:

1. Relevance: Everything we say to HCPs must be relevant and address their pain points.
2. Timing: The timing of message delivery must be at the point where they're already considering what you're offering.
3. Channel: We need to give HCPs the choice to use the channel they want to use to communicate with us.

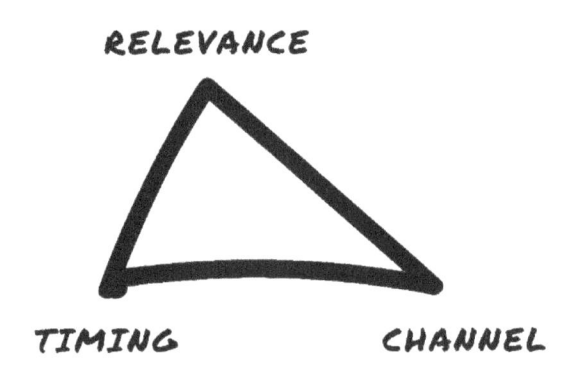

A triangle diagram with labels: RELEVANCE (top), TIMING (bottom left), CHANNEL (bottom right).

Many companies are only focused on the 'channel' element, and have started capturing HCPs' channel preferences to determine how to deliver information in the hopes that this alone will facilitate effective HCP engagement. But the truth is, HCPs' channel choices are not static and binary; they are fluid, depending on the type of information at hand and the interaction they have about it.

Also (to complicate matters further), when HCPs state their preferences, what they say may not necessarily be what they do. For example, a pharma leader once told me she was recently asked whether she preferred to go to the cinema or to stay at home and watch Netflix. She said she preferred the cinema, but when she looked at her pattern of behaviour in the past twelve months, she found she had not been to the cinema even once, but she *had* been watching Netflix on a regular basis. Therefore, when we ask customers about their preferences, what they tell us may differ from how they actually behave. That's where the data comes in: we need to measure their actual behaviours as well as their stated preferences.

We facilitated a joint workshop with Veeva in 2022, during which a pharma digital leader shared a survey they'd conducted to ascertain HCP channel preference in Middle Eastern countries. They ran an exercise using functionality that updated automatically once the HCP had finished the survey, reflecting the channel preference on the HCP's profile in the Veeva system. Once they finished the survey, they sent messages (with permission) via email, WhatsApp, and SMS, and they found the HCPs were acting differently to their stated preferences. They concluded that in order to reach the customer, you need to use a three-hundred-and-sixty-degree approach. Then, based on customer behaviour, they used an artificial intelligence tool that gives them information. For example, Dr X always opens his WhatsApp messages over the weekend and his emails during the working week (in the morning or evening), so based on this information (i.e., the HCP's actual behaviour), they can draw conclusions about their HCP's channel preference.

In a nutshell, to create personalised experiences for HCPs, we need to give HCPs the freedom to use the channels of their choice. This creates an opportunity for them to *want* to engage with us in a two-way communication, rather than us only having outbound communication with them.

Let me give you another example. In November 2022, I wanted to reach several senior leaders in the NHS responsible for healthcare budgets. These leaders make population-based decisions regarding formularies, policies, shared care protocols, and budgets. I emailed each of them with a concise message alluding to our previous conversation and inviting them to meet me so I could provide them with further information. I gave them three choices: let me know their availability and suggest a suitable time to meet, have a phone call with me, or use my electronic diary to book a Microsoft Teams meeting.

Eighty percent of them booked a thirty-minute Teams meeting with me using my electronic diary within forty-eight hours of me sending the email.

These very same leaders normally refrained from engaging with the industry, so why did they engage with me so quickly? Simple: because what I was offering to them was relevant and the timing was ideal (November—when I was contacting them—is when they start deciding on projects for the April of the following year). Plus, *they* chose the channel, not me.

As another example: a few years ago, I was promoting a respiratory brand and reached out to my network asking for the name of the lead respiratory pharmacist in an account I was targeting. One of my contacts gave me the HCP's name and added, 'Good luck. She will never see you!' This didn't deter me: I truly believe if we have something of value to say to HCPs, they are always open to giving us time. We just need to understand them and adapt to their preferences.

My first conversation with this pharmacist was very short, and she was certainly frosty, but I adapted my style and condensed the information I was giving to the key points, and no more. I respected and accepted her boundaries, and only contacted her when I had something of genuine interest to her.

It takes time to build trust, and over six years, I have had several telephone conversations with her, we have exchanged emails, and she has given me e-permission. She knows I am in the US and is flexible to meet me in the afternoons to fit with my work hours. During our last telephone conversation, I said I would contact her when I had something of value to share with her, and she said, 'I look forward to talking to you. I take your calls because you don't waste my time!'

Indulge me for a moment while I share a personal reflection on digital channels and emotional connection.

The other day, a Facebook post popped up on my feed that brought back memories of how Facebook played a significant role in reconnecting me and my now-husband in 2010. We had known each other back in Iran, but life had taken us in different directions, so we lost touch and ended up living in different parts of the world. However, thanks to Facebook, we found each other again and reignited our connection. As I looked at that post, I had a realisation about why I feel so comfortable using digital channels for engagement: it struck me that it's because I had once actively used these channels to maintain a long-distance relationship with my husband.

Our journey began with him in Australia and me in the UK. He at one time moved to Saudi for three years and then to the US while I resided in Scotland. Despite the distance and different time zones, we managed to stay connected and maintain a close relationship through various remote channels.

As a result, I've come to appreciate the power of digital channels in fostering emotional connections. From 2010 to 2016, I spent at least an hour a day using platforms like FaceTime, WhatsApp, Facebook Messenger, Telegram—you name it—keeping that relationship alive. Additionally, since 2017, I've worked remotely in the UK while living in the US, well before remote work became mainstream.

Maintaining this deep bond with someone I care about through consistent digital engagement for six years taught me invaluable lessons: it showed me that with digital channels, we can recreate the same chemistry and connection we experience when meeting face-to-face.

Why is this relevant to the issue we're seeing in pharma?

Well, simply: healthcare professionals are human beings, like all of us.

They read us just as we read them. So, when we communicate through remote channels, we must improve our ability to truly understand the other person by listening to their voice—its tone and tempo. To be authentic and genuine in our interactions, we need to feel the emotions we wish to convey, such as empathy, understanding and care. We can't fake it: when we express ourselves through digital channels, our emotions are projected through our voices and transmitted to the other person.

DT Consultancy's report for 2022[8] showed that human use of technology provides the best customer experience because it allows pharma to use the resulting data to connect with HCPs on their agenda and using their preferred channels, but on a human level. It enables field teams to be more personalised and to connect on an emotional level, and when field teams communicate a compelling story that evokes emotion, they motivate HCPs to listen and act for the greater good of their patients.

Before we move on, I want to share the pain point that HCPs don't often share with pharma professionals:

The pressure that is currently on HCPs is crippling. During the pandemic, I spoke with a friend of mine who worked as a charge nurse in theatres in the UK, and she told me it was impossible to deliver the care she had been trained to with such significant staff shortages. It was so stressful. Her role was clinical, not administrative, yet she was receiving hundreds of emails every day. How was she expected to deal with that? She couldn't do the admin during her clinical time while she was scrubbed up in theatre, so she ended up working long hours, under tremendous pressure, to compensate for staff shortages, and then dealing with a huge bundle of admin at the end of the day, when she should have been spending time relaxing at home. She has since retired and feels ten years younger! She said she would not go back even if they paid her five times the salary.

---

[8]DT Consultancy—*The State Of Customer Experience In The Pharmaceutical Industry*, 2022: HCP Interactions (Feb 2023).

This isn't a happy ending: the situation has become even more dire since she left. Her department has lost thirty-two experienced staff (approximately half of the staff force) in about eighteen months.

That is the reality of working in the NHS.

At the NEXT Pharma Summit in Dubrovnik, one particular session called Healing Our Healers left a lasting impression on me. The video and panel discussion touched me deeply. The panel talked about doctors' mental health and wellbeing, burnout, and bullying, and shared statistics on many who opt to leave the profession altogether or, worse, end their own life because of the overwhelming pressures. Dr Carys, a registrar working in the NHS, shared how pressure, lack of resources, and long hours impact doctors' and nurses' cognitive functions. She said doctors are trained to do risk/benefit analyses all the time, but with impaired cognitive function, the speed and quality of their decisions is affected.

I felt Dr Carys' pain, and it was not the first time I had heard of such a story. My son is a doctor, and as I listened, I worried about his mental health because he was (and still is) under tremendous pressure, similar to that which Dr Carys described. The pressure is overwhelming, and at the end of a tough day, doctors cannot even talk (or, indeed, complain) about it because of patient confidentiality.

It makes you wonder, while doctors and nurses care for others and put their own health at risk, who cares for them?

The panel encouraged us to care for our healers and to train pharma professionals as mental health first responders. It may be a while before this happens, but in the meantime, we can take an easy small step: stop wasting HCPs' time and only reach out to them when you have something of value to add that genuinely reduces their burden and addresses their pain points.

Checkpoint 4: Action Steps

- Check the DT Consultancy report referred to in this chapter to see where your company sits in relation to customer experience.

- Ensure the content you deliver to HCPs is relevant and genuinely merits them interrupting their patient consultations.
- Gather insight about HCPs to understand their pain points and time sensitive priorities.
- Understand that channel choice is only relevant if you have relevant content and know when the HCP needs information that addresses their pain point.
- To be customer centric you need to be on the same page as the customer, know their pain, and be clear about how your brands or services address their pain. Otherwise, you are just creating noise!

5
Raising the Tide Lifts All Boats

I alluded earlier that in 2022, Cheemia ran a workshop in collaboration with Veeva Systems in which we invited global leaders responsible for sales leadership, field force effectiveness, and marketing across Europe. What I didn't share was that the workshop was scheduled to run for ninety minutes, but most participants stayed on past that point because they were getting so much value from the peer-to-peer interactions they were having with colleagues in other companies.

There was a genuine buzz in the room, and for the first time, I felt a palpable desire amongst the attendees for collaboration across companies. Everyone was excited to be in an environment where they could speak openly about common challenges. Peers were sharing stories of struggles, initiatives they had tried, and successes, but most importantly, they were keen to define what 'good' looks like and how to recognise and reward it.

In addition to intercompany barriers, they recognised that we need to break down internal departmental silos between the sales, marketing, and medical departments to speed up transformation and become more customer centric. The internal silos could potentially hold us back or slow us down.

In another workshop in Zurich, similar issues were highlighted. The participants agreed that senior leaders are the ones who could play a significant role in breaking down the barriers and aligning the company behind one single goal and a shared vision, and to do this effectively, the senior leaders need to be able to tell good stories and explain the 'why'. The stories need to be compelling; to inspire everyone. And these stories need to be supported by investment and relevant KPIs that drive the outcomes and help nurture the desired behaviours across the whole organisation.

Collaborative working is not the goal; it is the vehicle that allows the organisation to *achieve* the goal. If each silo has its own objectives and those objectives are not aligned, the company could be pulling in separate directions.

Therefore, organisations must have common objectives that are shared across every department. This way, different functions within an organisation can come together and define what role they can play to achieve those objectives and share departmental needs and progress. Shared objectives are important because they drive the behaviours required to make the change.

They also need KPIs that measure and foster teamwork and cooperation so they can cohesively move forward.

Companies could deepen this understanding across the organisation by creating secondment opportunities in different departments. My career has been a zigzag of experiences, from nursing to working in various departments and functions in pharma and doing a postgraduate diploma in marketing and an MBA along the way, and the combination of these experiences broadened my understanding of business: different roles exposed me to different pressures and expectations and showed me the impact of what we do in the bigger picture.

In the same way that my transition from a nurse to a pharma sales professional left me with a deep empathy for and understanding of HCPs, working in different departments has enabled me to see the same issues from different perspectives. When I started working in the field again, I had a much better understanding of marketeers and medical colleagues, and I knew if they said no, it was not because they did not want to do it; it was because there were limitations that prevented them from doing it. With this understanding, I could listen, delay judgement, and connect with compassion and empathy. This understanding also enabled me to push back when it was justified.

In a similar way, during the workshop in Zurich in autumn 2022, I saw attendees from marketing, digital innovation, commercial, and medical working together to address the key challenges in pharma. It was reassuring to know that they were not alone in feeling the pressure; that their peers in other organisations were facing similar challenges. They were excited about working together to develop creative solutions, as an industry.

The fact was, individual companies could go and look for solutions alone, or they could break down the barriers and start sharing their best practices and work collaboratively across the industry. Now more so than at any other time in the past, there is a genuine desire to collaborate across companies; the problems we are facing are too big for one company to solve on its own.

Someone who shares the same passion and desire for collaboration is Florent Edouard, SVP, Global Head of Commercial Excellence at the Grünenthal Group. So, I invited Florent to share his perspective on this topic here:

Being a patient does not only mean being sick, diagnosed, treated, and hopefully cured; that would be by far too simple. Being a patient, whether a rare disease patient or a respiratory or diabetes patient suffering from chronic pain, often means being spun around in circles, going from 'wrong diagnosis' to 'bad choice of treatment', going through a frustrating repetition of symptom descriptions to broken file records, and sometimes seeing over fifty physicians over a multiyear journey, all without getting the right treatment. All patient advocates have hundreds of testimonials; anyone working with and talking to patients has heard and empathised with people who just wanted to have their life back.

Despite hundreds of billions being spent on improving healthcare, we are still not there, and while pharma companies do not have the accountability or the power to fix the healthcare systems around the world, we have in our hands everything we need to reduce entropy by helping to ensure that patients get the right treatment and that physicians are equipped with the right tools to diagnose, engage, and cure their patients.

But to get there, one of the first things we need to solve is the demand for HCPs to have a trustworthy healthcare partner that delivers relevant and good quality support for their daily practice. And in the last forty years, we have been very busy as an industry building a reputation of greedy businesses where profits are more important than health, and that has led to the pharma industry being ranked among the worst industries in terms of reputation, alongside oil producers and weapons vendors.

COVID-19 gave us a unique chance to make a comeback, and we seized it (some more than others) by rapidly developing reliable vaccines that would restart a world that had been brought to a standstill, and even for no profit, for the boldest among us. That improved our image and rating in the eyes of the general public, and we have a unique chance now to not let it go; to blow kindly on that little flame of hope that we could become what we should be: the reference point for when HCPs need information and support regarding the latest tools and medicines that will allow them to diagnose and cure their patients.

But to achieve this, we all need to realise that this customer experience—the image customers have of us and the trust they can put in us—relies on every single action we all do every day. It is not only related to what we do, but also to what our peers do. Any one of us can try to swim against the current, but if the pharma industry as a whole is not seen as trustworthy, we will not be able to progress against the accumulation of other efforts.

Because customers do not really differentiate us and despite all the pride and money we spend on expensive booths at congress and marketing efforts, most HCPs have a vague image of our companies, a more precise image of our products, and a very clear picture of their patients, because this is what matters to them. We are not Pfizer, AstraZeneca, or Sanofi to them; we are well-known medicine brands they recognise. And, as they are smart, they make us all believe they have a special relationship with each one of us. They manage us as a portfolio of providers.

So, our job, as disruptors, is to come back to what makes us a fantastic place to work. It is to go back to that patient—their journey and the pain and complexities that go with it—and look at anywhere we can improve it, lighten the burden, and deliver added value both to the patient and to the physician that is treating him or her.

It is also to join forces internally, across all functions—marketing, medical, sales, and digital—so we can offer our customers the tools they want and need, when and how they want them, so that they can engage their patients, discuss with them, understand their pain and worries, deliver against those, and ultimately achieve better clinical outcomes.

Only when we articulate a better customer experience and deliver against it will we be able to fully take our seat at the table as the trusted healthcare partners of the patients and the HCPs, and this will then result in a massive improvement of the delivered healthcare for all.

But are we ready to break down the silos? Are we ready to unlink tangible actions and expected financial returns by doing something just because it is the right thing to do?

How do we get this done? How do we create that sense of belonging in people who are usually incentivised and driven by their own companies? How can we get the ball rolling, starting from what the agile people call a Minimum Viable Product?

Well, there is one way, which is for us to make the gain and price worth the risk for the people who join us on the journey.

As an example, I want to share with you how we tried that with a group of likeminded people during the pandemic using the social media platform Clubhouse.

'Fancy a chat with Elon Musk? Or Kanye West?' Two immediate thoughts crossed my mind when my thirteen-year-old son suddenly asked this: *What would I say to those guys?* and, naturally, *What the heck are you talking about?* Without waiting for an answer, he continued, 'Get yourself an invite to Clubhouse, because that's what's happening.'

This was my personal introduction to a social media platform that is based on people getting together and talking, as equals, about both everything and nothing. As someone open-minded to new platforms, I was naturally intrigued enough to accept an invitation from serial health innovator and entrepreneur Tara Wacks. And so began a journey of me joining countless rooms where people excitedly felt like they'd got the golden ticket. Now, they could chat 'exclusively' about the rise of *digital health, startups, or any kind of topic you can possibly imagine, with people they'd never normally get to meet.*

Now, I am no healthcare spring chicken. I have spent many an hour in conference networking rooms, advisory boards, focus groups, and market research. So, imagine my surprise when I found that these conversations felt, suddenly, fresh. This was a new type of expression space, a set of unexpected conversations, and a noticeable step away from the traditional ways of discussing healthcare. It made an impression. I began to wonder how we could use this new media differently to change the game. But, like a new kid at school, I craved some familiarity at first. I searched for known faces and quickly happened upon two of my frequent 'disruptive' collaborators, Paul Simms and David Hunt, both of whom had joined the platform only a couple of weeks before and were also still finding their feet. We put our heads together and decided we could harness our collective influence to curate a new dialogue, unhindered by the conventional rules, where people from all walks of life could come and suggest how the life sciences could be transformed, with no money or glory involved.

As if to prove the hypothesis, we were quickly supported by passionate patient advocate Christine Von Raesfeld, who now leads many of our club's conversations despite never having worked in a pharma company. She embodies the idea that improving patient care can be more important than egos, employment status, and earnings per share.

Over the next few weeks, we evolved into a 'daily show' format. People are busy, so we agreed on a maximum of thirty minutes per session every day at the same time, so no scheduling was necessary. A tight time limit, but we agreed we'd always leave a few minutes spare for anyone to ask a question or throw in their perspective.

On Mondays, we tackled diversity and inclusion, a

longstanding issue in a white-heterosexual-male-dominated industry. Aurora Archer, Pamela Raitt, and Kelly Croce Sorg are leading a fascinating and provocative conversation about the new faces of our future industry.

Tuesdays and Wednesdays became days where we could talk about the creative ideas that are needed in order to revitalise our industry's communications and trust, and where we could hear about the career failures of executives in a constructive learning environment.

On Thursdays, we doubled down on the patient engagement and listening agenda, with the help of Christine and Kristof Vanfraechem, by giving them the mic to express their needs, hopes, and frustrations, and to propose solutions for patient empowerment.

Did I mention the refreshing nature of the conversation? Our no-cost, no-commitment, no-glory approach has opened unexpected doors, with very senior and distinguished leaders volunteering to take guest spots and the size of the club growing beyond one thousand five hundred members in just a few months.

Today, we feel like more of a movement than just a mere conversation on a trendy app. Our rooms are places where renegades converge naturally, where broken industry dialogues are restored, and where networking becomes about quality over quantity, joining executives, providers, patients, and even their families. Many report that they learn more in a thirty-minute session than in a whole month of more structured sessions.

This was made possible by a prior agreement that everyone talks as an equal, representing only oneself—and thus, a genuine and empathetic group of likeminded people, propelled by the same values and a desire to turn words into action, was born.

After some weeks, we saw a new initiative, bringing our club *out of Clubhouse and onto more established platforms such as Zoom. With no advertising, no communications campaign, and no marketing director, we registered more than one thousand people in mid-August.*

It turns out that a ragtag collective of wide-eyed industry optimists can create a stronger sense of engagement than even the most well-orchestrated, well-funded initiative, because we want to raise the tide for all boats. After all, death and taxes may be certainties, but so is the fact that we'll all become patients in our lifetime. This is surely what paying it forward means.

I am not going to linger on the destiny of Clubhouse and the possibility that it does not really deliver on its expectations. We have created a long-lasting community that now lives far beyond the original platform. My point is: it is on us. It is on each of us to start pushing the walls; to start reaching out to the movers and shakers that are everywhere in this industry but need to be helped to realise their full potential.

It is not a question of being a CEO or a sales rep, or a researcher in a lab. We are all part of an industry that delivers incredible, lifesaving benefits to patients who need it. Yes, we do not get it right every time. Yes, not everyone in the industry is a role model in terms of probity or morals. But that does not matter, because there are many, many who want to do the right thing for the patient, for the healthcare systems, and for the people who deliver treatments and care to patients. And that is our strength. If we unite beyond company borders and if we work relentlessly so that our main focus is benefitting the patients and caretakers, then we can change the game. Are you willing to join the movement?

Florent shares a compelling reason for us to work beyond company borders. Communicating and connecting with HCPs is an industry-wide challenge, and at the end of the day, we're only going to solve this by working together and by embracing both teams and technology. This will allow us to create the ultimate experience for HCPs.

In the cross-company collaborative workshop in Zurich, leaders from different companies worked together. We probed how to build success in the remote and digital world we're slowly becoming accustomed to, identified barriers we must overcome in order to achieve human transformation goals, and determined the best L&D processes and behavioural change techniques.

I'd like to share an insightful quote from the workshop: 'We're surrounded by process, and we build huge departments and teams, and when we see a gap, we put another process in place. But for me, I think what I've learned is it must be about collaboration, and you really cannot do that without that human factor.'

The large-scale upgrades we aspire to as we perform digital transformation will inevitably fail unless they begin at a human level. The companies who are taking their staff along the journey with them—empowering, enabling, and educating them along the way—are already seeing the best results, and in large part, this is due to increased investment into change management from the bottom up.

Unfortunately, there are still those who remain entrenched in legacy and act as change-blockers. Thus, we need to recognise that change is the driving force of innovation. This takes time, pressure, and a multidisciplinary approach. We must ask what our visions of the future for our people are and what lies in our path to reaching that future.

Checkpoint 5: Action Steps

- Remember silos are like cancer: you want to get rid of them systematically and proactively in order to have a healthy organisation and be customer centric.
- On a scale of one (meaning low) to ten (meaning high), score your

desire to break down internal and external silos.
- Consider: what have you done personally, or could you do personally, to close the gap?
- Consider: what drives your motivation to collaborate more?
- Consider: who advocates for collaboration in your organisation or in your professional network? How can you connect with them to share best practice and bounce ideas?
- Consider: what is one easy step you can take to bring the other departments closer together?

PART II

HOW DO WE GET OUT OF HERE?

6
Shifting Gears

Buckle up! We're now going to talk about how we're going to accelerate change.

Previously, we have explored 'why' the field teams are reluctant to embrace digital tools and tech. In this chapter, I want to focus more on 'what' organisations need to do and 'how' to engage the hearts and minds of the field teams in a way where they feel that the organisation and their managers understand them.

Henry Ford said, 'If you always do what you've always done, you'll always get what you've always got.' This is a simple yet powerful quote, and accordingly, pharma needs to do something different. This means that you (yes, *you*) need to look at this with a different mindset to that which you have employed thus far. Ask yourself these questions, and if you don't know the answers, you need to invest time to find them:

1. Do you understand the pain points and challenges your field team face?
2. How did you gather these insights?
3. How did you validate the severity of their pain?
4. How do you show to the field force that you understand them and their pain?

When I was a student nurse, I lived with a friend of mine, who was a trainee GP at the time. During the winter months, she was seeing many patients who had the flu daily, and was advising them to rest and stay hydrated, reassuring them they would be just fine. She expected them to just get on with it. It was not a big deal, after all.

Do you see the problem here? She was seeing so many patients every day that she was treating them on autopilot.

One day, I came home to find her lying on the sofa, looking pretty unwell. Since I was trained as a nurse, I checked to see how she was feeling and how this was affecting her. It turned out she had the flu: she was suffering from aches and pains and was totally congested. She felt very poorly indeed. In fact, she burst out, 'Please shoot me!' I chuckled and told her to stop being dramatic. She said, 'I did not realise how bad this could be. I will promise to be so much nicer to my patients from now on, and show empathy when they complain about their symptoms.'

Getting the flu helped her to connect with her patients and their suffering because she was now feeling it for herself.

Similarly, when your field teams are overwhelmed, scared, and uncomfortable with technology, telling them to 'be positive', 'get up', and 'get on with it' will create disconnect. To engage with them on a personal level, you need to first either experience their pain, or understand and connect to their pain and how this affects them, and acknowledge that it is not their fault that they feel this way.

Furthermore, the field teams' behavioural shift needs to start with a shift in mindset and a willingness to try, fail, learn, and try again. Henry Ford (also) said, 'Whether you think you can or you think you can't, you're right,' and that couldn't be truer for the field team's use of digital tools for engagement. Mindset matters, and confidence grows through trial-and-error.

I encourage the individuals and teams I manage to come up with new solutions. I tell them, 'Ask for forgiveness, not for permission,' because I want to encourage them to make decisions and grow in confidence. They often tell me that when they first hear this from me, they don't believe it;

only after they experience this or see someone else ask for forgiveness do they really believe that I am serious and that I mean it.

To create a learning culture, you need to create a safe environment that allows field teams to test, fail, and learn. To make this culture shift believable, you need to have real stories that support this vision, and you need first-line managers and middle managers to echo this ethos and support it.

If you want to teach your teenager to drive, you cannot expect to send them to an afternoon of driving lessons and them to become a competent driver by the time they get home. Using digital tools, like driving, is a practical skill that takes time for individuals to get their heads around. Some learn the concepts quicker than others, but generally speaking, it takes time to master the skills and become truly competent.

Changing behaviours requires effort and is not easy, but the first big step in the change process is understanding the 'why' across the organisation. The 'why' needs to be communicated by the leadership through compelling stories and narratives that evoke emotions and inspire everyone to unite behind it. Don't assume they already know it because you have communicated it once before; communication is not about the message that has been sent, but the message that has been received. The field team will change if there is a perceived direct benefit in doing so for them, so if this hasn't happened yet, you can guarantee that the team hasn't got the memo yet.

What motivated our team at Cheemia to use a mix of channels? Well, it started with us accepting the fact that HCPs' preferences have changed. The HCPs are already using multiple channels, so we just needed to understand the fact that if we increase our ability to reach key influential customers that impact formularies, policy, and guidelines through the use of a mix of channels, the team will hit the sales targets and get maximum sales bonuses. Moreover, we communicate with our client's marketing and medical teams regularly to articulate the different scenarios our field team encounters. We make suggestions for content creation for each scenario in mind of addressing the HCPs' pain points and challenges, and then marketing and medical go off and use this information to provide us with approved templates and content. Through this process, the field team feels valued, involved, and motivated to use the content immediately, because they have asked for it specifically, and in return for marketing's efforts, the field team provides immediate feedback to marketing.

The nature of the communication between the field team and marketing is agile, evolving, and focused on identifying challenges, testing and trying concepts, and figuring out what resonates with and engages HCPs. Marketing is not telling the field force what to do, and there is an adult-to-adult dialogue that involves articulating challenges and scenarios and developing solutions together. Marketing is asking the field team, 'What else do you need? How can we help you add value or address HCPs' challenges? What meetings should we invest in? Who are the key target audiences and target accounts that most benefit from this proposition?' and in turn the field professionals are giving valuable insight.

There is much that can be learned from this collaborative approach.

The ultimate nirvana for pharma is for it to be able to capture the same level of understanding and knowledge about HCPs that Facebook, Amazon, and Netflix have for their consumers. To achieve this, pharma needs to do something differently as an industry: leadership needs to communicate to the field teams the role the field teams must play in this journey if they are to reach deeper customer understanding.

Some pharma companies are taking a giant leap to empower field teams and move towards a customer centric approach by creating a single point of contact for HCPs in the field. Roche and Grünenthal are experimenting with new operating models to provide a single point of field contact to enhance the customer experience, and this hybrid model, in some cases, is a combination of KAM and MSL roles focusing on improving the customer relationship.

The field teams are most proficient at managing customer relationships, but they need to have the support of the whole organisation behind them if they are to become truly customer centric.

Here is what Paul Tunnah, Founder of pharmaphorum and, latterly,

Chief Content Officer and MD at Healthware Group, has to say about empowering the field force to be the heroes in the transformation story.

The question is: why haven't we achieved this already? After all, we have been talking about the concept of the orchestrator rep for at least a decade, but so few companies have managed to really join the dots in delivering on this.

In digital transformation, the 'PPT' framework is often referenced as a way of considering the key elements of change: People, Process, and Technology. It's been around since the 1960s, and the elements are, in my view, quite deliberately listed in that order, with the most important one (people) coming first.

The reality of life sciences and, indeed, many other industries is that companies try to drive change the other way round. Business leaders get excited by the new technology coming through, and the (often exaggerated—check out the Gartner Hype Cycle[9] for more on this) talk is about how it can revolutionise their results. So, they rush headlong into implementation programs that focus on rolling out this shiny

---

[9]www.gartner.com/en/research/methodologies/gartner-hype-cycle

new tech around the world. Then, the attention turns to the 'process' element as they rapidly craft new ways of working that fit around these new systems. Too often, the 'people' element amounts to simply passing on the initial sales pitch about how awesome this new tech is and how it's going to transform the company in an amazing way.

With this approach, the end users (in this case, the field teams) are often left wondering where all this change came from, why they should disrupt what they are already doing (which they believe is already working well), and, most critically, 'What's in it for me?' (WIIFM). If incentives are still aligned with previous ways of working and the associated KPIs, the answer in their mind to the WIIFM question is often 'not much'.

In order to shift this oft-repeated approach (which is doomed to failure from the outset), we need to apply exactly the same rigour and attention to the internal 'customers' of digital transformation, especially the field force, as we do to our engagement with external customers when trying to shift behaviour and prescribe habits around a new brand. We cannot just tell the field force the story of how exciting this change is going to be and expect them to take our word for it; we need to make the field force the central heroes in this story.

Before discussing how to do this, let's step back to consider the key elements of a good story. If you think about your favourite book or film, it's probably hard to put your finger on exactly why that's the case. Indeed, there are many elements to a good story, but some of the main ones are interesting characters, challenges, triumphs, and, perhaps most critically, experiences and emotions you can personally relate to (i.e., relevance).

For the field force to see themselves as the central characters—the heroes—of this story, we need to consider all these elements. From the outset, the key challenges that we are *trying to overcome (not just for the business, but for them as individuals) need to be communicated.*

The only way to do this is to gather their input at the start of the process, before designing and implementing the processes and technology. This not only has the advantage of securing critical input around where they see the challenges on the ground and what they need to overcome them, but also makes them feel part of the transformation (in this analogy, the story) and helps gain their buy-in, simply by engaging them upfront.

Once done, this also enables a much better articulation of the 'triumphs' that can be achieved both at the organisational level and for them as individuals. If we have on-the-ground input regarding what the challenges are, what the solutions might be, what tools they need to deliver success, and how we should be measuring success in this new world, then we're arming ourselves for a good outcome.

If you put this back into the context of external customer engagement, this is exactly the process we tend to follow. Think about the amount of time that is spent on HCP, payer, and patient research, message testing, advisory boards, insights into channel preferences, and so on. If we're willing to invest such time and money into the engagement approach for one brand, why not also do the same for an enterprise-level digital transformation that can drive better results for all brands?

Just like any brand campaign, this process doesn't stop once we've gathered input from the field force and launched the transformation. Instead, it needs to be an ongoing program of iterative feedback and adjustment, to ensure continued adaptation and improvement.

As I've outlined, input from the field force is a critical and often-overlooked element of success, but to really 'shift gears' and drive organisational change, they are not the only characters in the story. Ultimately, the field force is the frontline custodian of the relationship with the customers, so any approach has to work for these customers, too—that is, the HCPs, the payers, and, indeed, the patients.

Similarly, the multitude of 'central' roles involved in coordinating customer engagement, whether via the field force or directly, must also be consulted in the process—and this needs to be done across internal silos.

Am I suggesting that a commercial digital transformation aimed at empowering sales reps needs to also consult medical, corporate comms, compliance, and so forth? Absolutely, yes! For our external customers (HCPs, payers, and patients), the internal silos of a pharmaceutical company are meaningless. They simply want the right information at the right time, whether about the brand, a broader disease area, or the company. In the context of making the 'story' that we tell them meaningful, impactful, and relevant, they need to see a single, cohesive narrative, irrespective of whether it is told by the field force or home base, and regardless of which function is telling it.

To draw a direct parallel with the world of films, ask any Star Wars fan what they think of the eighth film in the series, The Last Jedi, and you will get very mixed opinions. When George Lucas wrote the original Star Wars (Episode IV: A New Hope), he had in mind the entire story arc for the first six films. Yet remarkably, when Disney came to develop films seven to nine, they didn't seem to have this, with different directors taking things in their own direction one film at a time. This is a rare mistake for Hollywood, but not for our world of pharmaceuticals, where the corporate, medical, and commercial stories don't always knit together so seamlessly.

Engage early, internally, externally, and across silos if you want to implement a system that tells a cohesive story to your own teams and your customers and that will deliver success.

Finally, every good film has an amazing trailer (and every

good book has a good blurb on the back cover). The most memorable movie trailers give us glimpses of the story that will unfold, often at the most dramatic moments, accompanied by music that makes the hairs on the back of your neck stand up, and sometimes (although this was a lot more prevalent pre-2000) a well-known voice narrating some key points. A good trailer builds the excitement ahead of the launch, and senior leaders need to do the same thing if they want to transform their field professionals: they need to convey early on what is coming, how we expect it to have an impact, and how everyone will be included in delivering its success. Early communication from leadership, including C-level, is needed to ensure the teams are aware of the 'what', 'why', and 'how'—and the 'WIIFM'. This is critical.

To come back to the start of this section: remember, it's People, Process, and Technology, in that order. The processes and technologies need be built around the desires and beliefs of the people they will serve, internally and externally. This is the missing piece of the puzzle. It's why at the industry level, we've not yet seen the promised transformation of field forces.

It's also more than 'change management', which is a term that's never sat quite right with me, perhaps because it implies forcing through change or dealing with inevitable resistance to it. But if the 'people' bit is done well, it's much more than that; it's 'change empowerment'—and it takes a village to make it happen, all helping to tell the same story.

If we continue doing what we have always done, we will not be able to create a personalised customer experience. So, we need to shift gears to achieve our shared vision for the future. This vision needs to be communicated in a way that evokes emotions, engages hearts and minds, and creates a supportive culture that cultivates change and growth across the organisation.

Leadership cannot do this alone. Field teams cannot do this alone. We are in this together.

Checkpoint 6: Action Steps

- Gather the field team's input at the start of the process, before designing and implementing the processes and technology. Gain their buy-in simply by engaging them upfront. If the field team give negative feedback once they put these tools to use, work with them to solve these issues.
- Create a learning culture in your organisation that encourages testing, failing, and learning. This culture needs to go beyond slogans and straplines, and needs to be supported by true stories from within the organisation.
- Ensure the learning culture is being sponsored and nurtured by senior management and first-line managers to support change of behaviours.
- Manage your expectations for omnichannel adoption. Allow more time for the field team to become competent.
- Empower the field team to articulate their challenges and define which customers and accounts are most likely to deliver outcomes.
- Value the field team's input for developing solutions. Listen to their suggestions and feedback to create content that is relevant to HCPs, and pay attention to the field team's experience with using the content.
- Explain why the field teams need to collect data and what role they play in building customer experience and how this is going to enhance their relationships with and access to HCPs. The field team's attention is focused on the customer, and because of that, they have built strong relationships with HCPs, so they will actively get involved in initiatives that bring more value to customers.

- Engage early, internally, externally, and across silos. Early communication from leadership (including the C-level) should focus on ensuring that the teams are aware of the 'what', 'why', and 'how'.
- Remember that it's 'People, Process, and Technology', in that order. The processes and technologies need to be built around the desires and beliefs of the people they will serve.

7

The Human Transformation of Digital Engagement: How Did We Do It?

———

Let us turn our attention to the human transformation of digital engagement and how to increase the adoption of digital channels amongst field teams in a way that inspires and motivates them to change their habits and behaviours.

Previously, we have covered why we need to shift gears and the fact that we need to do this together as an organisation. In this chapter, we will focus on the human element of digital transformation and how to overcome field professionals' resistance to adopting remote tools and digital content—in six weeks.

Yes, six weeks.

If you want to shift the behaviour of your field teams, you need to give them the practical tools they need in order to transition from being uncomfortable and unfamiliar with digital tools and digital content to being competent and confident in using them. Also, they need to trust that by using these tools, they reap benefits. These benefits could be the fact that they can fill their diary with appointments, engage healthcare professionals more effectively, and achieve their business objectives in a way where they get rewarded, get recognised financially and emotionally, and feel valued.

Our approach is unique because Cheemia ReSET was developed *by* a field team, *for* the field team.

Allow me to explain why this matters:

Cheemia ReSET, our award-winning online training platform, was designed by field professionals who are experts in remote selling. The main difference (compared to other training) is that the trainers not only understand the pain points and challenges faced by the field team, but they also have firsthand experience of it (think back to that example of the GP who then got the flu). They have gone through the transition themselves, and are therefore able to connect with field teams on a personal level.

What's more, the solutions are based on real-life experiences, not theoretical concepts. Marina, a sales manager in Spain, said she remembers the first time she watched the introduction video: 'The video was very personal, and I felt Mehrnaz was talking to me. I felt touched, and I felt already committed to the program.'

Often, the training organised by pharma companies or their external trainers is delivered in a group format, and the pace set is based on the lowest common denominator. This approach is counterproductive for the learner. Have you ever tried running with someone who has a much slower pace than you? It is so painful because you need to hold yourself back and cannot get into a proper rhythm. Similarly, when we learn, we all learn at a different pace and have different styles of learning, and trying to adapt to a different pace is excruciating at best and completely fruitless at worst.

I recall attending a two-day training session on a CRM system about ten years ago. We had to travel to England, and when we arrived, they gave us an inch-thick manual and slowly took us through it over two days. I was so bored in the training. I don't like reading manuals; I am an experimental learner. When I am learning new technology, I find it so much easier to watch a video showing me the steps than reading manuals. They could have condensed the two days of training into short, bite-sized videos for us to tap into on-demand!

Indeed, at Cheemia, we have found that, when it comes to practical skills training (especially for something you need to practice regularly), classroom-style learning does not suit individual preferences. In our experience, we have found training to be more engaging for the field force if it is in bite-sized video format, if it is user-led, and if it can be consumed at the user's preferred time and speed. This reflects the field team's preference, which is to have short, to-the-point, on-demand content, similar to the content pharma creates for healthcare professionals.

Thus, we broke down the field team's transformational journey into six essential milestones and developed a module for each, which covered everything in more detail. Each module had short video lessons accompanied by worksheets to encourage the field team to apply the learning to their own territory and daily practice. For example, Module 1 focused on getting the field teams into their high-performance zone and establishing the right mindset. In this module, we share our top secrets to help the field teams to get comfortable with blending face-to-face and alternative channels to engage HCPs and get amazing results. We also set them up for effective hybrid working and find the balance between living and working from home, without compromising either.

Let me share the introductory lesson from Module 1 here. This will allow you to see how and why this resonates with the field teams.

> We appreciate that reaching HCPs remotely can be difficult and, at times, challenging. While Lindsey and I chose to work remotely because of our personal circumstances, we recognise that this new way of hybrid or remote working probably wasn't your choice. We know that working remotely could lead to you feeling isolated and lonely, and we also understand that there are new channels and digital content and data that you (the field team) need to get your head around. All these tech tools can seem overwhelming and different to the way in which you worked before.
>
> We acknowledge that changing working practices and getting your arms around them can be challenging. To make

matters worse, pharma companies are still expecting performance, activity, and sales targets to be hit. Getting to grips with this new way of working at the same time as delivering the expected outcomes can be quite stressful. What's more, access to HCPs is more challenging, and there are compliance steps you need to take in order to gain e-permission and access via various channels.

There are days where you work really hard, spending all day in front of the screen or on the phone. You probably stay on hold for ages, and may even get through to the receptionists, but they (after all that work) probably just cut the conversation short, and you finish the day totally exhausted. At times, it may appear as though you've got nothing to show for all your efforts. You might be even feeling so stressed that you wake up in the middle of the night, or lay awake in your bed rehearsing what went on during the day, worrying about the next day and how that is going to pan out. You may ask yourself, How long this can go on for? How long can I cope with this level of stress and uncertainty?

We get that, because we felt it ourselves, and it's taken us years to get our heads around how to manage remote working in such a way where you can feel effective and engage HCPs. Accordingly, we've developed strategies, techniques, and templates that really helped us.

Over the next six modules, we want to share these with you so that you, too, can be in your 'high-performance zone', where you feel you can be productive, feel really good about yourself, and define what a good day would look and feel like.

When you get a commitment from a customer, a product listed on the formulary, or an agreement for a therapy review, you know you've had impact and influenced people. It feels great, right? Well, I want to show you how you can still feel energised; I want to help you get back to that spot. And yes, it is possible.

The whole point of this course is to get you to your high-performance zone so that at the end of the day, when you close down your computer, shut your office door, or put your work stuff away, you know you've done a really good job, and you get that sense of achievement and 'feel-good' factor.

The advice that we share with you is all based on our own practical experience: we ourselves are engaging HCPs by doing remote calls, using email, conducting video calls, and influencing HCPs on behalf of our clients. This is not something that we read in a textbook or some podcast that we are repackaging and telling you; this is based on our own collective experience over the last decade, between me, Lindsey, and the rest of the Cheemia team. All the advice we share has been tried and tested, and we have been able to achieve amazing results with this recipe. In some accounts, we've had single-digit market share, and by applying these principles, we've been able to increase the market share to sixty-five to seventy percent within twelve months. We want you to experience that same amazing feeling, get similar results, and help you be more confident and equipped to have effective hybrid engagement with your customers in both face-to-face and remote selling scenarios.

This course spans six modules. Here, we're going to focus on the first module, which concerns 'getting in the zone' and fired up for hybrid and remote working. We have five key pillars that help you to get to that stage, and today, we're going to share them with you one by one. So, let's see what those five key pillars are.

The first one is about the benefits of omnichannel selling. HCPs' preferences have shifted, and they are much more open to using remote channels for engagement. If you only rely on the face-face-channel, you are going to miss out on a lot of opportunities. Field teams who have applied the course content in their work have managed to increase their face-to-face and *digital engagement with HCPs by as much as five times.*

The second pillar that we want to share with you is something really important that has a big impact on your self-esteem and your wellbeing, and that's 'getting physically ready'. This is how you dress; how you turn up at work. Call it your work uniform. How you physically prepare and present yourself has a profound impact on the way you feel about yourself, your self-esteem, your self-confidence, and how you project yourself over the phone and online.

In the third lesson, we're going to talk about 'sorting your workstation' and how to prepare your work setup. We pay a lot of attention to this for two reasons. The first reason is because having the right environment supports your back, neck, and shoulders (so you don't cause yourself any damage while you work). The other reason is, it's so important to have a setup that reduces distractions and background noise, gives you the best possible audio-visual experience, and allows you to have your equipment ready (which will simplify your communications when you receive callbacks and get your diary filled up with appointments). Plus, if you don't have the right environment, that will affect your motivation, which could mean you miss out on engagement opportunities.

Lesson four is about the 'winning mindset'. Whether your thoughts are positive or negative has a huge impact on the way you interact and project yourself. We help you weed out those negative thoughts and rehearse the positive outcomes that you want to achieve so that you can get yourself into a productive 'winning mindset' and get the outcomes that you desire.

The final lesson is about movement, because when we move, we change our physiology, and this change in physiology changes our emotions. If you're sitting at your desk idle and are not getting up and moving, you'll be lucky if you hit two thousand steps by five o'clock. We want to draw your attention

> *to this and give you strategies to help you move more frequently and, in turn, maintain a high level of energy and increase your 'feel-good' factor. If you implement our strategies and tactics, you'll probably find that you end up being a lot fitter and more energised and motivated, and that positive feeling will spread to other aspects of your life, too.*

I hope this introduction gives you a flavour of the content for Module 1, as well as the tone and style that resonated with the field team.

Notably, we found that the things that triggered behavioural change lay not in the videos alone, but in the application of the knowledge via the worksheets we designed. It was 'doing the doing' that shifted the field team's behaviours, because when they put the learning into practice, they gained experience, and their confidence grew through this experimental learning and application.

The concepts we were asking them to apply were not rocket science. In many cases, there were things they knew already, but were (critically) not applying in practice. For example, field teams know they need to do pre-call planning, but they often do not do it, so we held our group accountable and encouraged them to do it—and as they started, they saw the impact immediately: they found that more HCPs were engaging with them and that the quality of their interactions had improved.

We also found that the individuals who were experiencing an improvement in the quantity and quality of HCP engagement started to organically share their positive results and success with their peers and their managers. This meant that the excitement about the platform was bubbling from the field teams, and their enthusiasm encouraged others to adopt it.

Koen Janssens, Associate Director of Field Force Effectiveness at Norgine, said at the beginning of the training, 'The field team's reaction was, "Here we go again! Another global training added to many we have already," but this was more special, as it touched the heart and soul of the people. They were at risk of losing their job, and Cheemia offered them a platform to continue doing their job.'

In parallel to Cheemia ReSET, we developed a management module called Cheemia ReINSPIRE. The management module gave the managers the tools they needed to coach their team on a concept they had little experience with themselves. Specifically, we provided first-line managers with the tools they'd need to hold the field teams accountable—to take massive imperfect action—and as they did, they found that transformation and change reliably took place, one individual at a time.

During the global rollout with pharma clients, we defined clear 'weekly milestones' for the field teams and field managers to accomplish. Certain weeks of the program were designed to focus on 'implementation' to ensure everyone had completed their worksheets and was applying the knowledge in practice. This way, everybody, across different countries, was moving in the same direction.

Upon reflection, we concluded that the role of the sales managers was fundamental in facilitating that change management.

The third key finding we took away from this was that confidence can grow really fast—in as little as six weeks. Through repeated practice, the team were making mistakes and learning from these experiences, and they were sharing their best practice with each other. The first-line managers were facilitating the sharing of best practice and were creating the learning culture in the team, but they were not the change agents: the field team were. As the field team went through the modules, week by week, their confidence grew through the application of the knowledge and practice. Each module built on the previous module, and by the time they completed Module 4, their confidence levels were increasing. By the time they were getting to Module 6, they were feeling so confident that they were coaching healthcare professionals on remote consultations.

People often assume digital transformation is easier for younger people, but given the right format and environment, *anybody* can change behaviours. The KAM who had the biggest improvement in embracing digital tools and content was fifty-nine at the time, which is amazing!

Overall, we found that long-term behavioural change is established when field professionals can learn from practical, in-field experiences, rather than classroom theory, and that the role of leadership is fundamental in facilitating this transformation. Perhaps most importantly, Cheemia ReSET was developed as a journey, not a one-off training. The field teams and first-line managers had access to the platform for twelve months so they could revisit and refresh on demand.

Here are the elements that we incorporated in our online training approach that resonated with the field teams and led to the rapid adoption of digital tools and content.

- The training was delivered by practicing remote sales engagement experts, not training experts.
- We created a six-module roadmap for the field teams to take them from feeling uncomfortable and unfamiliar to being confident.
- The user felt we were talking to them as an individual, rather than as a more generalised group.
- Each module contained several bite-sized videos that could be consumed at the user's pace and convenience, rather than one-off group training.
- We created engaging content that was reflective of the current reality field teams face, and provided practical tried-and-tested tips to overcome the challenges.
- We provided downloadable worksheets that helped the user put their training into use.
- Through the videos and management modules, we encouraged the users to try, fail, learn, and try again.
- The management module provided the managers with the tools they needed to assess, benchmark, mentor, and, most importantly, hold the field team accountable.
- The language and tone in the training was peer-to-peer style, and this led to increased engagement and a high completion rate.
- The training content showed an immediate return on investment. It not only increased remote engagement, but also face-to-face calls.

- The change in behaviour was sustained a year after the training had been completed, because the field teams could see the personal benefits they would gain and the increased quality of their engagements with HCPs.
- Successes were shared. The field teams learned from their peers and were encouraged to test and learn. This created a sense of community and a network of likeminded people.
- We provided ongoing access to the training for twelve months—not a one-off session where you are bombarded with information that gets forgotten quickly. Plus, each video focused on a single subject, so it was easy to find exactly what field teams wanted to revisit and refresh.
- We addressed mindset, motivation, habits, and behaviours, not just how to use the technology. This included managing emotions, dealing with rejection, and building resilience.
- We added quizzes and a certification to enhance the users' learning and sense of achievement.
- We shared how to overcome objections and barriers that stand in the way of field professionals engaging receptionists and secretaries. We also shared our strategies for reaching healthcare professionals and engaging them in conversations remotely.

What tangibly happens when you give the field team the right training to change their behaviours? Let's revisit the example we looked at in the Preface.

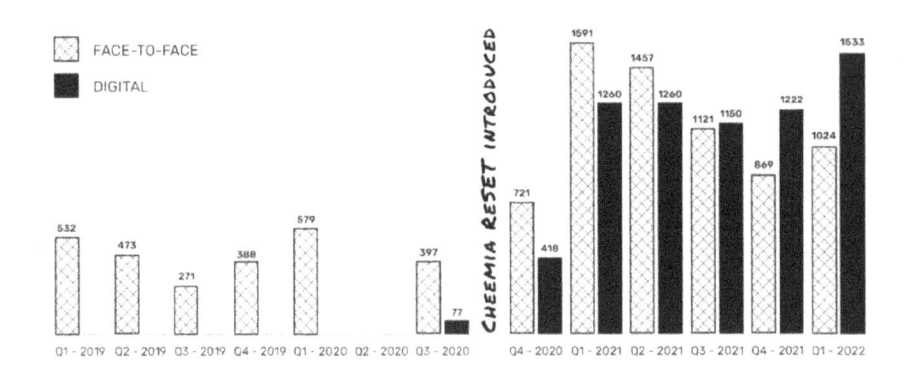

[Chart: Bar chart showing FACE-TO-FACE and DIGITAL interactions by quarter from Q1-2019 to Q1-2022, with "CHEEMIA RESET INTRODUCED" marked between Q3-2020 and Q4-2020. Values: Q1-2019: 532; Q2-2019: 473; Q3-2019: 271; Q4-2019: 388, 579; Q1-2020: 397; Q2-2020: 77; Q3-2020: 721, 418; Q4-2020: 1260, 1591; Q1-2021: 1260, 1457; Q2-2021: 1121, 1150; Q3-2021: 869, 1222; Q4-2021: 1024; Q1-2022: 1533.]

There was a significant increase in remote calls. We had expected this. You may recall, though, that the surprising observation was a dramatic increase in face-to-face interactions, too. Within six weeks of the training, the combination of face-to-face and digital contact with HCPs was significantly larger than the pre-pandemic call rates in Belgium (whose results are captured in this graph), Switzerland, and other European countries.

This is pretty remarkable: usually, when you do a training initiative, you see an initial change, but over time, people revert to their old habits. But this was not the case with Cheemia ReSET. Instead, we found that this increase in overall activity was still being maintained well over a year after the training had been completed.

This graph shows a fivefold increase in HCP engagement shortly after the training was finalised and a year after. I'm not someone who loves measuring coverage and frequency, but I am proud of what this graph indicates, as it shows the field teams' confidence in using digital channels. Naturally, you generate more activity and touchpoints when you connect with HCPs on their agenda, and this increase in engagement led to the company achieving their highest ever sales and profitability in 2021.

After I share these results with people, they often ask me, 'Did the company increase their field team in 2021?' and the answer is no; the results were achieved by the same individuals as before. They just got better at engaging with HCPs on their agenda and got more confident with using various channels in addition to face to face—and by doing that, they gained a significant advantage. Through Cheemia ReSET and Cheemia ReINSPIRE, we created a roadmap that allowed the field teams to adopt new skills, change their behaviours, and build their confidence. The field team's behaviour doesn't change because you tell them they must change; change happens when an individual understands the 'why' and accordingly decides they *want* to change. They also need to enjoy the experience of, and see the benefits of, change in order to continue. Then (and only then), change becomes their new norm and infiltrates their daily habits, and that is how the high level of HCP engagement you can see in the graph above was sustained.

I invited Koen Janssens to share his experience and perspective regarding how, by introducing Cheemia ReSET, they engaged hearts and minds and shifted behaviours. He shared the following.

> *Cheemia ReSET was our survival kit.*
>
> *The moment COVID broke out, some sales reps were thinking,* In one month's time, I might have to find another job. *To be honest, I felt fear and uncertainty in my role, too, as I was responsible for field force effectiveness. We were all sitting at home, not seeing customers anymore. We operate in the*

gastroenterology world, so for four to five months, everything came to a halt. Doctors could not physically accept patients in the hospital. All screening programmes stopped.

I think you can do two things. You can say, 'This is it, we can't do anything to change anything,' or you can be open to new things. Our mission is to bring medicines to patients and healthcare professionals, so we looked for any way to keep our business going. You can't say to a patient, 'Sorry, there is no way that we can contact doctors, so you will have to wait another year.'

We started to realise that we needed to train people on remote engagement—and believe me, if you are in the middle of the ocean and someone throws you a buoy, then you must grab it. Accordingly, I was quite inspired, convinced, and passionate about Cheemia ReSET, and I managed to transfer that into the organisation. We went to our fields teams and said, 'Listen, guys, we know this is a tough time, but we want to invest in you,' and we saw in the first pilot that everybody was behind this immediately, as it was seen as their ticket to picking up the relationship with their doctors again.

In the beginning, people said, 'It seems very simple.' Some of the stuff in the modules is not rocket science. But let's be honest, sales is not rocket science. It's introduction, questioning, presentation, handling objections, and closing. But if you were to do a very critical analysis and evaluate yourself—record yourself and review it later—you would see how many mistakes you make or how often you forget about these core things.

Once they started doing it, they realised that it was not so simple. They realised it needed practice. It needed rehearsal. It needed preparation. You have to train every day, just like Messi does; just like Ronaldo does. They're brilliant football players, but they're still on the pitch three or four times a week, training.

As they went through the program at their pace, they started to implement, step-by-step, something from every module. When connecting for the first time, they would do try-out sessions with people they knew or with doctors who they have a good relationship with. So, it was in a very safe environment where they could even have a laugh when it went wrong, without feeling that it would be seen as unprofessional. What we saw was people quickly gained confidence in using the video channels and engaging remotely.

We saw them sharing their successes with each other, and to encourage this further, we used peers to expand on the good practices, not management. For example, if someone was good at great openings, they would lead peer-to-peer exercises for that module, or if someone was good at overcoming the big objections or the recurrent objections of the gatekeepers, they would take the lead with that module.

In parallel, the field teams also forced marketing to think about rapidly developing digital content, because the more remote engagement they had, the more content they wanted to be available to share. So, digital content was accelerated because of the demand of the salesforce, not the demand of marketing.

I think that a lot of people have also been inspired internally. We see marketing has joined and MSLs are joining, as these people need to be trained in these skills, too.

Then, we started to think about how management could coach people remotely. This is where Cheemia ReINSPIRE helped. By giving them the tools, the first-line managers were able to coach and assess people through the training program.

We have a saying here: if you forget to prepare yourself, you prepare yourself to be forgotten. This is truer than ever with remote engagement, because if the doctor feels you have not prepared at all, then the likelihood of having follow-up sessions will be lower, because the doctor sees this as a waste of time. If, *however, the doctor really feels you took the time to prepare, look at his case, and see what historically has been said, and you build on that, then we see some people very easily getting more remote appointments.*

I always say to the people on this remote engagement training that if you use ten percent, well done: try to add another ten percent. There's always room for improvement. If you use that for all your goals and you can convince every HCP to see two or three hidden patients he was not treating with your product, and if you do that systematically and for all your HCPs, this is an exponential return on investment.

By doing remote work well, there was a boost in face-to-face meetings, too. We saw that the doctors were very open to planning again. Besides our normal face-to-face frequency (three or four times a year), they also allowed us to contact them remotely. I think from the perception of a doctor, they don't see remote as a physical meeting. If you add all these remote channels together, you have contacted that doctor more than ten times. If you saw them in person ten times, they would say this was overkill, but they allow you to go in person a few times, do video calls a couple of times, and call them and send them emails many times, because the communication comes from different angles. And that's why we ended up with more frequency than before.

We know companies that still struggle with access due to the fact that they didn't handle the whole remote journey very well and were incapable of or unskilled at remote engagement. For example, instead of gently opening a conversation and building a rapport with the HCP, they would immediately start to pitch their product. These are the companies who were regularly refused access because the HCPs would feel that, as a human being, they were not really being respected, and were only thought of as the facilitator of a commercial relationship with the pharma company.

The whole thing with omnichannel is the profile of a doctor. The challenge for pharma companies is knowing how to get and keep the attention of each individual doctor, and omnichannel is the answer: not every doctor is responsive to face-to-face, so if you offer all these channels, some are more responsive to an email, and some are more responsive to a video call. You need to capture the data regarding which channel each doctor is most responsive to. Then again, we need to be careful not to burn it immediately. If you know a doctor is responsive to emails—that is, he reads, opens them, and clicks through from them—that doesn't mean you should send them an email every week (which some companies do). In this situation, it will not take long for them to withdraw their consent because you are abusing it.

We started Cheemia ReSET at end of 2020. 2021 was a very good year for us. Profit, turnover… all the records were smashed in 2021. Even 2022 was a good year for Norgine, even though there were other problems due to the economic crisis, pressure from governments on the healthcare industries and regulations, and some products being taken out of reimbursement. But we didn't feel that, because once the value of your product has been made clear, the doctor is already convinced by the added value.

As we move on, we see also that the new generation of HCPs expects this to happen. They don't want to have face-to-face meetings all the time. And in that situation, it's very important to be able to reproduce the face-to-face chemistry remotely, because if the doctor sees this as quite artificial or robotic, they will think, *I can read a document on the Internet, or, I can watch a video, or, I can ask for an extract from congress.* The more that remote engagement comes close to face-to-face chemistry, the more a HCP sees this as a trusted advisor that adds value. And the ones that have difficulties with it need to train and retrain themselves.

*

The takeaway message from this chapter is: field training needs to change behaviours. Just because you've completed training and ticked that box doesn't mean the training has necessarily had an impact.

Checkpoint 7: Action Steps

- Leaders should clearly articulate the benefits and importance of omnichannel to the field teams.
- Redesign your training style and approach to align with the preferences of the field team who are on the go. Provide bite-sized, to-the-point, and on-demand content, similar to the content that is created for healthcare professionals.
- Use language that makes the field team feel you are talking to them as an individual.
- Provide training and resources to the field teams that resonate with the reality of the work they do and are based on real field practical solutions.
- Ensure training is ongoing and mapped as a journey.
- Encourage the field teams to apply the learning in practice.

In summary, you need to offer a comprehensive roadmap and resources to the field team that empower them to learn and adopt omnichannel practices. We offer this kind of training through Cheemia ReSET and Cheemia ReINSPIRE. Specifically, we:

- Provide first-line managers with bespoke tools to support them in coaching the field team and holding them accountable in 'the doing'.
- Foster a learning culture. Creating an environment that encourages continuous learning and development is a must, so we teach and embody the idea that it is okay to try, test, fail, and learn.

- Encourage autonomy and ownership. Allowing team members to have a certain level of autonomy and decision-making authority can increase their motivation.

8

From Message Bearers to Trusted Advisors

Let's now look at how field teams could make the leap from merely delivering marketing messages to becoming trusted advisors for healthcare professionals.

Some believe the jump to becoming a trusted advisor for HCPs is a leap too far. I, however, argue that it is more than possible; we just need to take decisive steps in this direction and be bold and brave.

Stephen Covey, the author of *The 7 Habits of Highly Effective People*, says, 'Seek to understand before you are understood.' Do we do this in pharma? Perhaps not! More often, we rock up with our sales materials and dive in, irrespective of the HCP's level of interest. We look for every opportunity to flash our three key marketing messages.

Typical pharma.

Our KPIs measure share of voice, coverage, and frequency. We don't have KPIs to measure the quality of the relationships we build, or the value we bring to HCPs. This is a problem: to become trusted advisors, we need to provide sound advice that adds value in every touchpoint. If we do this consistently over time, we can become the HCPs' trusted advisors.

Hence, to achieve this, we need to shift our engagement model with healthcare professionals from one that prioritises merely delivering sales messages three or four times a year (regardless of their quality), to one that focuses on adding value through every touchpoint (no matter how many of those there are).

In Chapter 4, I shared three elements that are required for effective HCP engagement in the form of a triangle—relevance, timing, channel—and you need to keep this in mind with respect to HCPs. To deliver value to HCPs, we need to have a deeper understanding of them: their personas, desires, and goals. How often do we genuinely stop and try to see the world from their perspective? If I challenged you to write a SWOT analysis from the customer's perspective, how easy would you find this task?

Some pharma companies have started adding customer centric SWOT analyses to their business planning process to bring them closer understanding of their customers, but when it comes to writing them, they find they don't know enough about their customers, and they revert to write a brand centric SWOT analysis.

The value being given to HCPs needs to be built into the core of our communications, not put as a wrapper around them. If our communication addresses HCPs' unmet needs when they need it, they *will* engage.

Thus, to become a trusted advisor, we need to be brave and bold and focus on HCPs' needs.

> *Robert Louis Stevenson's quote ('Everyone lives by selling something') suggests that everyone, in some way or another, is engaged in selling, whether a product, service, or even themselves, and it is true: you are always selling an idea, product, or service. Teachers are selling to students, parents are selling to children, children are selling to their parents. A professional salesperson understands that selling is about doing something for someone, not to someone, and so the professional salesperson will take the time to find out as much about the client that they can, get permission to ask them questions, and*

find out what they want to do and accomplish. In essence, they get on the same mental frequency as their prospect, and they genuinely help the prospect to get where they want to go. Empathy equals balance.[10]

Understanding HCPs allows us to see the world from their perspective and get on the same page, and that is when they trust us—and rightly so! Customer centricity is not about doing something *to* our customers, it is all about doing something *for* them, in line with their needs and preferences.

As we discuss this, it is necessary that I bring some home truths and bust some myths. We need to accept the reality, after all.

1. Don't Expect Your Brand to Be Relevant to Every HCP

Field teams need the analytical capability to segment and identify the HCPs and accounts who are best matched to the brand proposition, filter this down further to those who are actively looking for a solution, and deliver the brand propositions that meet the prospect's needs, directly or indirectly. It's all about identifying the twenty percent of accounts and customers who have the potential to contribute eighty percent of sales.

2. Not All Locations Have Equal Potential for Brand Growth

Market dynamics in some locations and healthcare systems are better aligned with the brand proposition than others, and understanding these dynamics can lead to better alignment with healthcare providers' priorities

---

[10]Proctor, B. (2015): *The Law of Vibration—The Secret Law of Attraction Coaching*. Shared on YouTube.

and the creation of joint initiatives and long-term trust at a country or regional level. Be bold and focus on the locations that can deliver more value.

In a similar vein, instead of dividing our resources equally across every territory, we need to make differential investments based on account potential, customer needs, and customer desire to work with us. This requires a different mindset—one that is willing to take calculated risks.

3. Don't Waste HCPs' Time

Instead of driving coverage and frequency, we should only engage with HCPs when we have something of value to share with them—that is, something relevant to them. This means calling on fewer customers, taking more time to research their needs and networks, and exploring various channels (direct and indirect) to reach them. We find this approach leads to the hyper-personalisation of communications, which leads to greater engagement with HCPs and greater trust. It also allows us to reach HCPs who don't normally engage with pharma, because they know we don't waste their time.

4. HCPs Don't Always Know What They Want

When Apple created the iPhone, no one said, 'I want an iPhone,' or (maybe more accurately), 'I want a smartphone in my pocket,' but now, everyone wants one. Similarly, HCPs don't always know what they want, but it is our job to create solutions that create a positive experience, nonetheless. Keep this simple and easy from a usability perspective, and from a value perspective, create FOMO (fear of missing out): make it special to the point where HCPs are thinking, *I need to join these events because I'm learning a lot and they're valuable to me.*

HCPs hate it when we ask them to tell us what they need, because they don't know how to describe something they have not seen. So, take a concept to them and let them shape it. Collaborating with HCPs and involving them in the process gives them the chance to know us and our intentions, and is a quick way to build trust.

5. HCPs Prefer to Have a Human Point of Contact

When HCPs have a question and need help, they prefer to reach out to a human, either via phone or email. What's more, they prefer to have a single point of contact—preferably someone they have met in person or virtually who is responsive to their requests and can channel them to the relevant person in the company to get the best answers as quickly as possible.

6. Field Teams Need to Be Channel Fluid

Instead of overusing the channels the field team prefer, we should be asking HCPs how *they* would like to have a conversation. Would they like us to send an email and give them the content beforehand? Would they like us to see them face-to-face? Would they like us to engage with them remotely? It's about being flexible and finding out what is right for that particular HCP in that moment.

7. Don't Expect All HCPs to Be Tech Savvy

One of the leaders who attended our workshop and was an agile program manager (they had been recruited from outside the pharma industry) shared this story with us:

> *I had an opportunity to do a field visit with one of the sales teams*

to attend some of the meetings they have with the HCPs, and it was fascinating; eye-opening. I asked HCPs about their openness to digital platforms, and their response was, 'We know that this is the way the world is moving—we know that this is where digital and technological advancement is progressing and evolving towards—but I'm a slow typer, some of the systems take ages to set up, and we have some senior HCPs who are struggling to use the technology because they can't remember their credentials and don't know how to log in, so they need some help from their team members.'

Although most HCPs are leaning towards digital, there are some who struggle with it, so pharma needs to be flexible and engage based on HCPs' preferences, whether digital or otherwise.

The Power of Human Connection

Building relationships requires people skills, and field teams have the natural ability to connect with others. This is especially important to note when you consider the fact that people skills are not something you can teach from scratch: you can develop them further, but you need to have the natural ability to connect with people in the first place in order to really excel at them. Those who are able to easily connect with others are naturally good at listening and responding appropriately on the spot.

Certain skills around engagement, negotiations, and influencing people are naturally present in 'people'-people, and they can be nurtured and further developed. Without these human skills, organisations won't have the ability to influence HCPs by evoking emotion and inspiring change. Many experienced field professionals have invested in and built relationships with HCPs over time, and these relationships facilitate access and give insight into their environment and priorities. I, for one, have been working remotely from the US for the last seven years and have been able to engage with senior leaders in the NHS in locations where I have not worked before. What's more, I find that the time it takes me to build relationships and become their trusted advisor is much shorter now compared to when I worked in the field seeing them face-to-face, and the main reason for this is that we use remote channels to reach them and clearly communicate how we can help them and add value.

Another reason for this rapid engagement is, health services generally operate in silos. Secondary care is disconnected from primary care, and these silos don't effectively connect to payers and other aligned services, such as mental health service providers. In order to reach agreements regarding inclusions for formularies, you need to be able to break down the silos and join the dots. This is where field teams can have a huge impact that bots cannot: a skilled key account manager can help HCPs to navigate their way across the fragmented healthcare system silos, get the relevant project and protocols off the ground, achieve their desired outcomes, and improve patient access to medicines. Payers and heads of medicine management are aware of their internal silos and often find it challenging to effectively communicate policy and guidelines across the integrated care system, so now, we have the capability within pharma to support healthcare providers by working collaboratively and complementing their communication cascade. This not only supports the communication of their policy and improves patient access, but helps to grow brand market share dramatically in a short period of time.

A recent survey of four hundred and fifty pharmaceutical companies[11] showed that only four percent of digital-only launches have been successful. Clearly, we need to use a mix of human and digital to reach 'trusted advisor' status.

Think about where you go to get advice and who your trusted advisors are when it comes to buying pharma services and products. Do you reach out to your usual vendors and people who you have known and trusted for

---

[11]Iskowitz, M. (2022): *Built to fail: Pharma's digital product launches plagued by single-digit success rates*. Published in *Medical Marketing and Media*.

a while? Or are you open to talking to new vendors—people you don't know who send you unsolicited emails or reach out to you via LinkedIn asking to meet you for fifteen or twenty minutes to solve your challenges?

People buy from people and are influenced by their social network. When we book hotels or Airbnbs, we check rating scores by other travellers. When we buy products, we read user reviews and recommendations. HCPs are the same: they make decisions based on their personal experiences and the experiences of those they trust, or whose experiences they think could reflect theirs. So, consider: who are HCPs' trusted advisors? They may initially search online or visit a trusted website, networking hub, specialist publication, or even an AI tool like ChatGPT, but because of information overload and digital fatigue, they'll always reach out to another human being to draw meaning from all this information.

Field teams have the potential to be that human being that the HCPs trust and reach out to for help and advice. What the HCPs want is an easy, accessible source of information they can tap into on demand, along with a human face that can help them navigate their way to the answers they are looking for.

I met an inspiring Executive Director of Customer Facing Field Teams from the US at the NEXT Pharma Summit in Dubrovnik. He gave a presentation on the field team's journey from sales reps to trusted HCP advisors. I loved this presentation—it was music to my ears!—so I asked him to contribute to this chapter and share his expertise and observations.

> *COVID-19 took its toll on healthcare in more ways than we know. Many experienced and renowned physicians retired early, and lot of office staff, nurse practitioners, and physician assistants left their jobs. As a result, sales reps lost critical access to many physicians, impeding HCPs' education on modern medicine—education that is critical to get patients on therapy faster. In this great reset, to extend patients' lives and improve the quality of care, sales reps and pharma companies*

must meet the physicians/customers where they are going on this journey with patients.

While many physicians have adopted telemedicine to care for patients, sales reps still greatly rely on face-to-face customer engagement, with little to no virtual engagement.

We must understand that HCP engagement is constantly evolving.

- Hybrid is the new standard. More than fifty percent of reps in leading companies now do roughly two virtual calls plus eight emails per week (reps of the future use their Fridays to schedule approved emails).
- Remote beats face-to-face. There is twice the number of promotional responses from virtual calls than from face-to-face calls.
- Content drives the impact. There was a threefold increase in content use in virtual meetings.
- HCPs need real-time service. More than seventy-five percent of HCPs need help to process information with real-time services.

In summary, if customer engagement is not insights-driven and personalised, HCPs see it as noise. In our experience, a trusted HCP advisor is a digitally enabled sales rep, who uses content more frequently (five times more frequently) and conducts more HCP meetings (forty percent more). This leads to more patient treatment starts (sixty percent more).

In order to elevate the reps of today through the lens of them being trusted HCP advisors, many myths need to face reality. To give you an idea, I am listing a few below:

- A rep of today will assume that Next Best Action will eventually replace sales reps. However, a trusted HCP advisor knows that Next Best Action capabilities are there to complement and enhance sales reps' hard work, keeping

- *them ahead of the competition, not to replace them.*
- *A rep of today will assume that Next Best Action is a set of multichannel tactics that sales reps already utilise. However, a trusted HCP advisor understands that Next Best Action facilitates omnichannel with orchestrated, personalised, and insights-driven engagements.*
- *A rep of today will assume that dismissing the CRM suggestions will reflect poorly on them. However, a trusted HCP advisor understands that it is critical to either approve or dismiss CRM suggestions to increase their future accuracy. This allows for a better understanding of what is and isn't resonating with customers.*
- *A rep of today fears capturing authentic HCP email addresses in CRM, for they could be used for unintended marketing campaigns. However, a trusted HCP advisor demands strong consent management practices that ensure that field intel is compliantly captured and utilised by sales for in person, and not misused for non-promotional reach.*
- *A rep of today fears that recording a negative sentiment to key messages in CRM will reflect poorly on their performance. However, a trusted HCP advisor understands that sentiments reflect how certain marketing messaging landed with individual HCPs. They realise that not all messages will land positively, and that's okay!*
- *A rep of today sees no opportunity for seamless handover among fellow customer facing teams. However, a trusted HCP advisor understands that an effective CRM is a single and shared CRM where all teams (commercial, medical, market access, and marketing) compliantly coexist and can seamlessly 'pass the baton' to serve customers on their journey.*
- *A rep of today believes that Next Best Action is limited to*

> *approved emails only. However, a trusted HCP advisor understands that the future of Next Best Action is HCPs experiencing optimisation through next best sample, next best key message, account plans, banners, web traffic, and appropriate cross-team collaboration.*
>
> *In summary, a trusted HCP advisor understands that not all working days go according to the plan. To serve at the point of care, they need to connect, schedule, meet, execute, and share to deliver a concierge service.*
> *Across all marketing channels, trusted HCP advisors make the biggest contribution to impactable sales for a reason: they provide on-demand, real-time, customer centric services.*
> *The rep of the future understands that 'if sales is asking for a date with the HCP, marketing is the reason why the HCP would say yes'.*

Aligning with the customer's agenda is the fastest route to becoming a trusted advisor, and this can only be done by enabling field teams to confidently connect and support healthcare professionals with data driven insights. It is also the fastest route to moving the brand proposition forward, gaining formularies, influencing policies, and driving the adoption of the product in question.

This brave and bold mindset shift must happen across all departments.

Field teams have the capability to become trusted advisors for the marketing and medical departments, thereby creating two-way communication, where field insight is shared with marketing and content for HCPs is co-created with input from the field team. This may mean having fewer field professionals, but those who remain in the field will need to be more proficient and competent to a level that allows them to become that trusted advisor, both internally and externally. The Omni Advantage occurs when field teams use their excellent 'people' skills, become tech and data savvy, are confident in using various channels, bring insight to medical and marketing departments, and help HCPs to achieve their goals.

Let us move to the next chapter to examine the habits, skills, and competencies that are required in order for field professionals to operate at this level.

Checkpoint 8: Action Steps

- To become a trusted advisor, you need to be brave and bold and focus on HCPs' needs.
- Ensure customer engagement is insights-driven and personalised.
- Remember that being 'customer centric' means doing something for the customer, not to the customer.
- Identify and focus on the HCPs and accounts that benefit most from the brand proposition in question.
- Don't waste HCPs' time! Make sure you add value in every touchpoint.
- Take concepts to HCPs and shape them with their input and involvement.
- Provide HCPs with a human point of contact who can triage their requests.
- Ensure that your field team is competent and can efficiently and flexibly use remote channels as well as engage in face-to-face meetings.
- Understand that it takes time and consistency for trust to be built. Invest in this process!

9
The Future Field Teams

In this chapter, we will examine future pharma field teams' habits, skills, and competencies—that is, the things they will have done and will continue to do so they can maximise on the Omni Advantage and be empowered to become the heroes in this transformation story.

I believe the field team's role is now more important than ever, because going forward, we're going to have fewer field teams and those who remain need to be highly skilled; to be trusted advisors for healthcare professionals and internally, for the marketing and leadership teams. Accordingly, we need to treat the field teams as adults and give them accountability—and with that accountability comes the responsibility to deliver and behave in a new way.

In marketing, we have the seven Ps, but here, I have produced nine Cs to illustrate some of the essential skills and habits required for field teams to effectively connect with HCPs. The nine Cs are as follows.

1. Connect

Field forces need to communicate confidently, both face-to-face and remotely, to a level where they can create real chemistry and emotion and tell a really good story that takes the healthcare professional on a journey. If this is done well, by the time they've finished the conversation, they will have not only conveyed key messages, but they will have created a memorable emotion and feeling in the healthcare professional. After this, the HCP should be motivated and compelled to take action.

2. Close

The future pharma rep needs to close with action. I have interviewed and observed many senior key account managers, and one of the biggest gaps I see in their roleplay is they don't close properly—that is, they don't define the next steps, such as when they're going to meet next, what the customer is going to do as a result of this interaction, and how they're going to move the project forward. Even if the customer says 'yes', the field professional needs to be really clear about what that 'yes' means in terms of next steps. Who else is going to be involved in their project? When are they going to do what they said they were going to do, and how is delivery going to be measured? If we agree on a true win-win agenda with HCPs, there's no reason why they wouldn't want to commit to taking action, right?

The underlying reason for field teams not closing is often lack of proper call planning and a lousy call opening: they don't communicate a compelling reason for why that healthcare professional should listen to them, and without a compelling reason that addresses the healthcare professional's pain point and connects with them on their agenda, the sales call turns into a tell sell. The HCP will switch off, and may end the call early.

This brings us to the third C, which is…

3. Collect (Insights)

This is about asking open questions when interacting with healthcare professionals and really listening. I've spoken to many healthcare professionals who say that generally, pharma field teams don't listen: they have their talking points, and they listen to the HCP just enough to find a gap to talk about their own talking points. They ask questions, but their responses are not connected to the information that has been conveyed by the healthcare professional.

Thus, we need to encourage the field team to listen and to seek to understand the healthcare professionals, rather than just focusing on conveying the key messages. Perhaps some of this is our own fault, since we give the field teams three key messages that we expect them to deliver to the healthcare professionals during their interactions. In every roleplay exercise or field visit, we measure the delivery of these key messages through specific KPIs, and this puts the field team under pressure to deliver on these KPIs. It's like a bullshit bingo: they need to get these words across somewhere in the conversation, and that pressure leads to the field professionals not concentrating on the insights they should be collecting.

I think we need to take a step back, really listen, and measure call effectiveness by the quality of the insight that's being collected, the quality of the responses to the questions, and the information that is being gathered by the field team about the healthcare professionals.

4. (Be) Credible

The future field teams need to know their stuff—and by that, I mean clinical scientific data across sales and MSL teams. When we recruit field professionals, they are educated to degree level and have an interest in science, and/or were previously healthcare professionals. Therefore, they've got the capability to understand clinical information and communicate it. To be credible, we need to invest in ongoing clinical training to ensure the field professionals know their clinical scientific data inside out and can hold a conversation with a specialist at their level.

5. Crunch (Data)

This is the ability to effectively analyse data, do bottom-up forecasting, determine exactly where the uplift will come from (that is, which accounts and which customers), segment the customers and accounts, and do effective targeting. If field professionals take accountability for targeting and sales, they're going to be much more focused on picking the targets that are going to feasibly deliver the needed outcome. If, on the other hand, the organisation or marketing department selects these targets for the field professionals and sets arbitrary coverage and frequency goals, those goals will be achieved, but the outcomes will probably not be delivered.

We need to move away from measuring surrogate markers and treat field teams as adults. We need to give them accountability; to allow them to define what needs to happen to deliver those outcomes. Because at the end of the day, if they are achieving the coverage and frequency goals that have been given but they are not delivering the outcomes, they're going to come back and say, 'Well, you told me to see these people. I could have told you that was not going to deliver the outcome.' But if we allow them to choose the target, frequency, and forecast, the field team will be laser-focused on delivering these outcomes, and as they gain experience through trial-and-error, they'll learn and become competent in targeting and bottom-up forecasting.

6. Collaborate

Collaborating means being a team player and internally aligning with colleagues to meet the healthcare professionals' needs. This requires a mindset shift, from a 'flying solo', territorial headspace to a mindset that prioritises cooperation across the wider team and that views the medical and marketing departments and internal colleagues as allies. The field professionals need to proactively inform the other departments and mobilise internal functions to better meet the healthcare professionals' needs.

7. (Be) Curious

Learning needs to be fluid, and to achieve this, field teams need to be curious and proactive, but not dependent on being spoon fed. This primarily means having a 'can-do' mindset. The right mindset is so fundamental in driving change that if the field professional has this, they will automatically approach problems differently: they will see them as challenges that they're going to overcome rather than walls that are going to stop them in their tracks, and if they're curious, they're going to find different ways of solving these problems and different sources of information that they can piece together to come up with creative solutions.

8. Competent (At Prioritisation)

This is a key skill gap. Many field professionals are good at planning and have had comprehensive training around business planning—they are able to do SWOT analyses, identify the key issues, and create detailed tactical plans—but they don't necessarily have the ability, knowledge, or skills for prioritisation.

Being competent at prioritisation is undoubtedly more important than planning, because if you want to have *x* number of units in sales, you need to prioritise your accounts based on opportunity and your ability to compete in that market first. Once you identify the accounts that are collectively able to give you the growth you want, *then* you start planning for those.

In our experience here at Cheemia, we have found that we can grow brands exponentially by focusing on a small number of accounts. Across the UK, we gain our growth from twenty percent of accounts, and those accounts give exponential growth that led to increased sales and market share in a relatively short period of time.

My point? Less is more and brings focus, and prioritisation is the essential yeast of significant exponential growth.

9. Confident (In Tech)

This means being confident in using omnichannel tools and tech. Confidence comes with practice, so we need to create a culture that encourages field teams to 'try, test, fail, learn, and try again'. Investing in the right training and giving them the step-by-step roadmap to mastering these tools will accelerate the adoption of these tools and, in turn, the field professionals' confidence in using them.

Many pharma companies have developed new competency frameworks for digital transformation to facilitate this change, and most of these competencies are filled with buzzwords—which makes the whole process even more complicated. The competencies don't come with a blueprint to explain what the field teams need to do to achieve them. It is like giving somebody a picture of a meal and asking them to recreate it without giving them the recipe or the ingredients. Companies need to break down the competencies into habits and behaviours, show the field teams how to get their heads around them, and provide examples of what that looks like and how to apply it every day at work.

To develop a new habit, many think you need motivation and willpower, but in my view, motivation and willpower will only take you so far. This is because building new habits is difficult if you don't follow the basic principles of forming new habits: making the habit obvious, attractive, easy, and satisfying.

To encourage your field teams to develop new habits and change their behaviours, you first need to demystify the process and give them a simple roadmap to adopting these new habits and skills. This will, in turn, make the habit obvious, easy, attractive, and satisfying.

To help you along this journey, we have created a roadmap for field teams. Specifically, this roadmap details seven habits field professionals must adopt in order to gain the Omni Advantage. If you are interested to find out more and experience one of these lessons, head to www.theomniadvantage.info/readapt-preview, or scan the QR code below.

Checkpoint 9: Action Steps

- Improve your field team's skills to a level where they can connect emotionally with HCPs to create impact.
- Invest in 'closing' skills training.
- Encourage your field team to listen and respond based on what the HCPs have shared and collect insight.
- Your field team needs to be highly competent in their product and therapy knowledge, so validate their competency in this area regularly.
- Improve your field team's ability to analyse data and expect them

to know their numbers.
- Foster a culture of collaboration and reward this behaviour.
- Reward and recognise curiosity amongst the field team (to encourage it).
- Develop the field team's prioritisation skills; this will have more impact than planning to grow sales.
- Build your field team's confidence in using tech using the guidance in this book.

10

The 'Doing' Chapter

In this chapter, I encourage you to roll up your sleeves and take massive imperfect action. After all, what is the point of reading this book and finding the gaps in your organisation if you are not going to do anything about them?

As previously mentioned, here at Cheemia, we commissioned research in 2022. The study involved several in-depth interviews with pharmaceutical leaders who were responsible for commercial excellence, marketing, and medical functions across the industry. The conclusion from this report indicated that transformation would not happen until the people within the organisation had themselves been transformed.

Here are the conclusions we drew from the research and the ten steps outlined for achieving change.

1. Recognise

We're not going back to 'before', even if that were possible, because customers in most territories *want* to change. This edict comes from leadership.

2. Set (Realistic) Goals

Change is not something that happens overnight. It requires investing in change management. It requires time, pressure, and a multidisciplinary approach. (It's hard to measure ROI in the moment, but the companies that are investing in these things seem to be seeing the best results.)

3. Expect Resistance

People don't want to change, so you need to work on the 'why' and the mindset before the specifics. Only after that can you focus on channels, behaviours, content, and so on. We tend to rush too quickly into omnichannel management, so be wary of doing this.

4. Vary Your Story

Produce multiple 'angles' for people to get their 'A-Ha!' moments in different ways. Some people will be driven by the recognition that there are better ways to provide customer value; others are fearful of the progress of others; some simply care about cost. You need to appeal to all these different motivators.

5. Unify Customer Engagement

Often, this means increasing the (historically lower) scale and agency of medical. All companies are converging. Some are focusing on cross-functional structures; others are combining a 'single' customer experience.

6. Personalisation Is Possible

...and should be worked on to ensure that customers are learning at their current level of understanding, in their preferred channel, at their preferred time...

7. ...But Requires Data IQ

...by improving capabilities in data handling and finding insights (e.g., through social listening). Everyone needs to be good at this, not just specialists.

8. Content Must Scale

...and it must do so in both quality and quantity, with dedicated content strategy and collaborative production. Digital distribution enables easy replication.

9. Invest in the Front End

The field professional in question is only one channel, but will continue to be an important one, even if just digitally. Accordingly, field professionals need to be great communicators, content facilitators, and storytellers, to an internal audience as well as external stakeholders.

10. Allocate Budget for Front End Training

Most marketing budgets are controlled at global level, but the global teams are not responsible for execution. In the same way companies allocate budget for digital content creation and platforms, you also need to plan and budget for people development at the global and/or affiliate level and assign accountability for its execution. This investment is perhaps more important than content creation, because it ensures that the digital content and tools are used by front end teams in HCP engagement.

Checkpoint 10: Action Steps

- Download 'Gaining Omni Advantage — Implementation Plan for Pharma Leaders' via www.theomniadvantage.info/implementation-planning-tool or the below QR code and populate it for your organisation.
- Complete Step 1: Assessment: Define Where You Are Now.
- Complete Step 2: Goal Setting: Define Where You Want to Be.
- Complete Step 3: Gap Analysis: Define What Needs to Happen to Achieve Your Goal.
- Complete Step 4: Leadership Accountability: Define Leadership's Role to Drive It.
- Complete Step 5: Training Design: Understand and Align with User Personas.
- Complete Step 6: Implementation Plan: Achieve the Journey. Assign a budget and agree on KPIs to measure ROI.

Postface
Moving Forward

It was June 2003 at a Pfizer sales conference that I met Bear Grylls. He was the keynote speaker, and he was talking about a time when he'd climbed Everest.

Months before this feat, while he'd been lying in his bed recovering from a broken back, Bear had stared at a poster of Mount Everest on his wall, and that was when he decided he was going to climb it. At that time, he was not even able to stand or walk, but he set himself a goal, and he was determined to achieve it.

I was really inspired by his story, and I thought to myself, *I want to go to experience what he's experienced.* As I was listening to him, I decided there and then I was going to hike to Everest base camp—and I knew I was going to do it. I didn't know exactly when, but the goal was set, and it was going to happen.

A month later, I was in London visiting my dear friend Debbie. Debbie is passionate about adventure holidays, and had already hiked to Mount Kilimanjaro, so I said to her, 'You know what, Debbie? I met this incredible guy, Bear Grylls, and since then I have read his book, and I really want to go to the base camp at Everest. I don't know when, but I really want to go. I wondered if you'd be interested in going together at some point?'

Debbie looked at me calmly and said, 'I'm going there in October. Do you want to join me?'

She shared her trip notes with me, and it sounded truly amazing, so I went back home and spoke to my family and my employer to seek their support and get the time off.

Before I knew it, the day was upon us and, to this day, hiking to base camp is probably the hardest thing I've ever done. Along that journey, I learned the biggest lesson of my life.

It took us seventeen days of climbing up and down, and while we were there, we also climbed Kala Patthar (which is slightly higher, at an altitude of five thousand six hundred and forty-four meters). I'll never forget the day we walked towards base camp. When you are at a high altitude, you can hardly breathe, your body slows down, and every cell in your body says to you, 'You should just turn around and go back.' The oxygen level drops, your mind gets fuzzy, and you become delirious. But I was determined and knew the only thing I had to do on that day was reach that goal; to just take one step at a time; to put one foot in front of the other.

Walking to base camp at Everest is quite strange. It is not going up a mountain and back again. You can see it for miles, yet you walk for ages, and you feel as if the journey is never ending.

I remember that, when we were halfway there, Debbie reminded me, 'You know, we're going there, but we've got to come all the way back.'

'Debbie, let's not think about coming back, especially not at this moment,' I said. I was lightheaded and struggling to keep my normal walking pace, and what kept me on that path was simplicity: just putting one foot in front of the other and moving forward.

That's the biggest lesson I have learned in life: to achieve a goal, no matter how big or how difficult it is, all we need to do is just take one step at a time and keep moving forward.

In this book, my focus has been on helping you accelerate behavioural change amongst your field teams to adopt omnichannel and digital content for HCP engagement. In order to keep moving forward (which is a must, whether you're climbing a mountain or creating impact in your industry), you need to:

- Measure where you are on this journey.
- Define where you want to be in one year, three years, and five years. Involve your team in drafting SMART objectives for this vision.
- Do a gap analysis to define two to three priority areas to focus on. Remember, less is more!

Change happens when you, as a leader, take action. So, don't just put this book down and carry on as normal. Start your plan today. If you are serious about change, start by completing the action steps listed in this book's chapters (if you haven't already), lead by example, and ask yourself what you will do differently today, tomorrow, and moving forward.

Also remember that changing behaviours is a journey. You must create a culture to develop new skills, support innovative ideas, and embed new habits. As a change leader, you can set an example and pave the way by taking action, trying something new, learning from failures, and celebrating successes. Be bold and communicate that you are not seeking perfection, but progress and innovation.

Be a role model for change and you will not only engage your field teams' hearts and minds, but you will also accelerate behavioural change in your company and gain the Omni Advantage.

Acknowledgements

Transformation is a journey and, accordingly, this book is designed to help you navigate your way along your own digital transformational journey. In my pursuit to explore and understand the human element of transformation (a pursuit which led to me writing this book), I had the counsel and input of several inspiring leaders in the pharmaceutical industry.

I would like to thank Paul Simms, Chief Executive of Impatient Health, for his continuous support and for acting as a sounding board.

I would like to thank Florent Edouard SVP, Global Head of Commercial Excellence, Grünenthal; Koen Janssens, Associate Director Field Force Effectiveness at Norgine; James Harper, Founder and Managing Director at twentyeightb; and Paul Tunnah, Founder of pharmaphorum and, latterly, Chief Content Officer and MD at Healthware Group; for their input and contributions to some of these chapters.

I would also like to thank the Cheemia team and our clients, who have field tested the concepts shared in this book.

And last but not least, I would like to thank Giles Etherington and Sophie Robertson, who have supported me in mapping out and editing these chapters.

Travelling along this transformational journey with other inspiring leaders and entrepreneurs has made it a fun and enjoyable adventure. We debated and bounced ideas around and together created something that we hope will shine a light on your path to gaining the Omni Advantage.

I welcome the opportunity to engage with readers of this book, fostering a continued exchange of valuable insights and best practices. I invite you to connect with me on LinkedIn to further these discussions, and for those interested in exploring further, I encourage you to explore our acclaimed training platform, Cheemia ReSET, by visiting www.cheemiareset.com. Let's connect and collaborate to enhance our understanding of omnichannel engagement in pharma. After all, we are all in this together.

About the Author

Mehrnaz Campbell is the Founder and CEO of Cheemia, a Scottish company that is managed remotely from the US. Mehrnaz has a successful track record of gaining market access and exponentially growing brands on behalf of pharma companies in the UK while assisting the NHS in achieving millions in efficiency savings and improving patient outcomes.

Born in Iran, Mehrnaz relocated to the UK at the age of twenty-one to pursue higher education and her career ambitions. Her passion for learning is reflected in her academic qualifications, which encompass general nursing, a postgraduate diploma in management studies, a postgraduate diploma in marketing, and a master's in business administration.

With over thirty-seven years of experience in UK healthcare, Mehrnaz started as a nurse and transitioned to various sales and marketing leadership positions with pharmaceutical giants such as GSK, Pfizer, and Takeda. Her final role was as Multichannel Strategy Director at Takeda UK before embarking on her entrepreneurial journey by founding Cheemia in 2017.

In August 2020, she introduced Cheemia ReSET, a multi-award-winning digital transformation platform developed during lockdown. This platform empowers pharma field teams to confidently utilise digital channels to deliver value to healthcare professionals (HCPs). The online training platform has not only helped global sales teams survive, but also thrive, achieving outstanding sales results, earning bonuses, advancing careers, and increasing earnings per share for pharmaceutical companies.

www.cheemiareset.com

Resources

Cheemia ReSET/Veeva
Workshop Report
www.theomniadvantage.info/
report

Sneak Preview of
Cheemia ReADAPT
www.theomniadvantage.info/
readapt-preview

Implementation Planning Tool
www.theomniadvantage.info/
implementation-planning-tool

Mehrnaz Campbell's
LinkedIn Profile
www.linkedin.com/
in/mehrnazcampbell